新一代信息技术系列教材

# 基于新信息技术的计算机英语

主　编　黄利红　周海珍　杨爱武
副主编　熊登峰　左向荣　曾　琴
　　　　郭俊兰　马瑶琴

西安电子科技大学出版社

## 内 容 简 介

本书面向高职高专计算机及相关专业学生，全面介绍了计算机专业的基础英语知识，内容涉及计算机发展史、计算机硬件、操作系统、应用软件、编程语言、互联网、信息安全等主题。

全书共 7 个单元，每单元均分为情景对话、阅读材料和扩展阅读三部分，旨在提高学生在学习和职场环境下的英语会话水平。本书从词汇、句法和结构等方面入手，加强学生专业文献读、译和写方面的基本应用技巧，并补充计算机小知识阅读材料，培养学生用英文解决计算机相关问题的能力。

本书可作为国内各类高职高专院校信息技术、计算机应用、计算机信息管理、软件技术、网络管理等专业的教学用书，也可作为 IT 行业技术人员和计算机爱好者的参考用书。

### 图书在版编目(CIP)数据

**基于新信息技术的计算机英语** / 黄利红，周海珍，杨爱武主编. —西安：
西安电子科技大学出版社，2019.6
ISBN 978-7-5606-5272-6

Ⅰ. ①基… Ⅱ. ①黄… ②周… ③杨… Ⅲ. ①电子计算机—英语 Ⅳ. ①TP3

### 中国版本图书馆 CIP 数据核字(2019)第 034798 号

| | |
|---|---|
| 策划编辑 | 杨丕勇 |
| 责任编辑 | 徐忆红 |
| 出版发行 | 西安电子科技大学出版社(西安市太白南路 2 号) |
| 电　　话 | (029)88242885　88201467　　邮　编　710071 |
| 网　　址 | www.xduph.com　　　　电子邮箱　xdupfxb001@163.com |
| 经　　销 | 新华书店 |
| 印刷单位 | 咸阳华盛印务有限责任公司 |
| 版　　次 | 2019 年 6 月第 1 版　　2019 年 6 月第 1 次印刷 |
| 开　　本 | 787 毫米×960 毫米　1/16　　印　张　9 |
| 字　　数 | 154 千字 |
| 印　　数 | 1～3000 册 |
| 定　　价 | 26.00 元 |

ISBN 978-7-5606-5272-6 / TP

XDUP 5574001-1

\*\*\*如有印装问题可调换\*\*\*

# 前　　言

计算机专业英语属于专门用途英语，是计算机及计算机相关专业的专业基础课。教育部高教司高职处与全国高职高专大学英语教育指导委员会明确规定，专门用途英语在教学中必须强调以应用能力为主线的思想，以应用为目的，以够用为度，以实用为主，重视培养实际使用英语进行交际的能力。

本书的主要特色如下：

(1) 切合学生实际，内容形式多样，在编写过程中注重学生的接受能力，这些主要体现在篇幅和语言表达上。

(2) 内容覆盖面尽可能广。为使学生在学习英语的同时了解计算机应用领域的基础知识，本书精选了来自英文教程、杂志或网站等的相关资料，主要包括知识概述、应用、专业分析、科技新闻、问题解决、心得交流等方面的内容。

(3) 在组织形式上，每单元引入相关的情景对话，培养学生在生活、学习和工作中运用英语进行专业交际的能力；加强计算机专业英语基本的读、译、写能力，目的是让学生在掌握基本必备知识的前提下学会举一反三、自主学习。

全书共 7 个单元，内容围绕计算机发展史、计算机硬件、操作系统、应用软件、编程语言、互联网、信息安全等方面展开。每单元均由情景对话、阅读材料和扩展阅读组成。

本书由黄利红、周海珍、杨爱武任主编，熊登峰、左向荣、曾琴、郭俊兰、马瑶琴任副主编。黄利红编写了单元1、单元2和附录，周海珍编写了单元3，杨爱武编写了单元4，左向荣编写了单元5，曾琴、郭俊兰编写了单元6，熊登峰编写了单元7。

由于编者水平有限，加上时间仓促，书中难免有不妥之处，恳请广大读者批评指正，提出宝贵意见和建议。编者邮箱为48735920@qq.com。

编　者
2018年12月

# Contents

**Unit 1　An Overview of Computers** .................................................................................................. 1
　　Section 1　Situational Dialogue：Computer Debate .................................................................. 1
　　Section 2　Reading Material: Instruction to the Development of Computers ............................ 2
　　Section 3　Reading Material: Classification of Computers Systems .......................................... 6
　　Section 4　Extended Reading: What are the Benefits of Computers in School? ...................... 10
　　Section 5　Extended Reading: What are the Advantages of Computers in Business? ............. 11
　　Section 6　Extended Reading: Commencement Speech at Stanford Given by Steve Jobs ...... 13

**Unit 2　Computer Hardware** .......................................................................................................... 19
　　Section 1　Situational Dialogue：Teens and Computers .......................................................... 19
　　Section 2　Reading Material: The Hardware ........................................................................... 20
　　Section 3　Reading Material: The Motherboard ...................................................................... 25
　　Section 4　Extended Reading: How Bluetooth is Making Our Lives Easier ........................... 28
　　Section 5　Extended Reading: Alibaba Group will Invest $15B into a New Global Research and
　　　　　　　Development Program .......................................................................................... 29
　　Section 6　Extended Reading: 10 Ways to Make a Computer Move Faster ............................ 30

**Unit 3　Operating System** ............................................................................................................... 33
　　Section 1　Situational Dialogue：System Crash ...................................................................... 33
　　Section 2　Reading Material: Operating System ..................................................................... 35
　　Section 3　Reading Materia: An Overview on the Most Commonly-used OS ........................ 39

| Section 4 | Extended Reading: Android Operating System | 43 |
| Section 5 | Extended Reading: Comparison between Linux and Windows | 44 |
| Section 6 | Extended Reading: Should You Upgrade or Repair Your Computer? | 46 |

## Unit 4  Application Software ... 49

| Section 1 | Situational Dialogue: A Job Interview | 49 |
| Section 2 | Reading Material: Instruction to the Computer Application Software | 51 |
| Section 3 | Reading Material: Using the Speak Feature in Office 2010 | 53 |
| Section 4 | Extended Reading: The Office 2010 Suite will be Released | 58 |
| Section 5 | Extended Reading: Microsoft Teams Challenges Slack for Office Dominance | 60 |
| Section 6 | Extended Reading: The Way Ahead: Innovating Together in China—by William H. Gates | 61 |

## Unit 5  Programming Language ... 65

| Section 1 | Situational Dialogue: Call about Courses of Computer Programming | 65 |
| Section 2 | Reading Material: Introduction to Programming Language | 66 |
| Section 3 | Reading Material: The C Programming Language | 69 |
| Section 4 | Extended Reading: What is the "Java language"? | 73 |
| Section 5 | Extended Reading: Object-oriented Programming | 74 |
| Section 6 | Extended Reading: How to Learn a Programming Language | 76 |

## Unit 6  Network and Internet ... 83

| Section 1 | Situational Dialogue: About Going Online | 83 |
| Section 2 | Reading Material: Instruction to the Network | 84 |
| Section 3 | Reading Material: Instruction to the Internet | 87 |
| Section 4 | Extended Reading: How to Use the Internet | 91 |
| Section 5 | Extended Reading: Set up an FTP between Two Computers | 95 |
| Section 6 | Extended Reading: To Be Completely Anonymous on the Internet | 101 |

## Unit 7　Information Security ........................................................................................................ 104

Section 1　Situational Dialogue：Computer Hackers ................................................................ 104

Section 2　Reading Material: Computer Viruses ....................................................................... 105

Section 3　Reading Material: The Ways to Protect Information Security ................................. 109

Section 4　Extended Reading：What Does a Data Security Manager Do? ............................... 112

Section 5　Extended Reading：How to Create a Secure and Stable Windows System ............. 113

Section 6　Extended Reading：Create a Perfect Password: Ten Easy Steps to Stay Secure ..... 114

## Appendix　Index of Basic Vocabulary ................................................................................... 118
## References ................................................................................................................................ 136

# Unit 1

## An Overview of Computers

After reading this unit and completing the exercises, you will be able to
- Be familiar with the history of computer development.
- Identify all the general elements of a computer structure.
- Apply your knowledge when using your computers.

### ☞ Section 1  Situational Dialogue:

## Computer Debate

Marion: Hey, Todd, you are exactly the person I am looking for. I have a favor to ask you.

Todd: Sure.

Marion: Well, the thing is, I'm thinking of buying a computer. OK, but I really don't know too much about computers, you know that, so I want to know, which is better, a laptop or a desktop? Which do you think I should buy?

Todd: Ah, that's a good question. Well, I'm not a computer pro but, I guess if you get a desktop computer for your house there's a lot of advantages. It has more memory and more power. You can get a bigger monitor for watching movies and things like that, and I think the number one reason that I like the desktop is it doesn't break. And you know it can't be stolen.

Marion: As easily, yeah.

Todd: Right, so when you have the laptop, you can take it and carry it everywhere. That means it's easier to break, you know you can drop it (yeah) or it's easy to be stolen, so yeah. On the other hand, if you just want to be mobile with your computer then obviously you want a laptop because you can take it everywhere, and you can use it in different places, so that's actually why I have two. I have a desktop and a laptop.

Marion: So you have the best of both worlds, really.

Todd: So what do you want to use your computer for?

Marion: Well, mainly I want to use it to keep in touch with my families and friends. That's my number one reason. I'd also like to use it for work, to make work sheets and that kind of thing for school, and the other thing is, oh yes, I want to use it to store my photos from my digital camera. That's really important to me because I'm afraid that I'm going to lose one of the CD's and all of those photos or something like that.

Todd: Well, how often do you think you'll be taking a computer with you, to work or somewhere?

Marion: I don't know really. I mean, I suppose I could just buy a laptop and leave it at home, couldn't I?

Todd: Sure. Sure. Yeah, I guess if you're just going to use for basic stuff then maybe, a laptop is good for you.

Marion: Yeah. Yeah, maybe. It's so confusing. It's such a big decision for me. I don't know what to do. Thanks for your advice.

Todd: Sure.

## ☞ Section 2  Reading Material:

# Instruction to the Development of Computers

Modern development in computers was started in Cambridge, England, by Charles Babbage, a mathematics professor. He began to design an automatic mechanical calculating machine called a difference engine, but in 1833 he lost interest because he thought he had a better idea — the construction of a fully program-controlled automatic mechanical digital computer. Babbage called this idea an Analytical Engine. The ideas of this design showed a lot of foresight, although this couldn't be appreciated until a full century later.

**The first generation**

The first generation of computers is generally considered to include machines built between 1946 and 1959, of which the ENIAC (the Electronic Numerical Integrator and Computer) was the prototype. ENIAC was built by two professors at the University of Pennsylvania in 1946. It included 18,000 vacuum tubes, weighed more than 30 tons, occupied 15,000 square feet of floor space, and consumed about 180,000 watts of electrical power. The ENIAC could perform 5,000 additions or 500 multiplications per minute.

In the early 1950s, the first mass-produced machines became available. The IBM 650, introduced in 1954, was the first commercially successful computer.

The first generation of computers was characterized by the use of vacuum tubes and regenerative capacitor memories. These expensive and bulky computers used machine language for computing and could solve just one problem at a time. They did not support multitasking.

**The second generation**

The second generation computers employed a new technological innovation: the transistor. In 1956, the transistors were first used in the building of computers. In the 1960s, transistor-based computers replaced vacuum tubes. Transistors had numerous advantages over vacuum tubes. They were smaller, cheaper, and gave off less heat.The second generation computers used magnetic cores as their primary memory. They used punched cards for input and assembly language. These computers gave users a significant increase in available memory (about 20x). Calculation speeds also increased.

IBM dominated the market of the second generation. Two of IBM's product lines were especially successful: the large 7000-series, and the small 1400-series.

**The third generation**

On April 7, 1964, IBM released its System/360 line of computers. The System/360's release marked the beginning of the third generation of computers. The System/360 computers used integrated circuits rather than individual transistors. This increased the speed and efficiency of computers. Operating systems were the human interface to computing operations, keyboards and monitors became the input-output devices. Magnetic core memory was replaced with semiconductor memories.

The notion of upward compatibility was introduced during the third generation. (This means that applications made for a given computer/system will work with the next model, just like an Excel 97 spreadsheet will work with Excel 2000.) Sophisticated operating systems were introduced, giving used unprecedented control over the computer.

IBM's System/360 and System/370 dominated the third generation computer market through the 1970s.

**The fourth generation**

Changes after the IBM System/360 were evolutionary, building on existing technology rather than completely replacing existing technology. Introduction of the microprocessors (thousands of integrated circuits placed onto a silicon chip) was the hallmark of the fourth generation computers.

In the 1980's, Very Large Scale Integration (VLSI), in which hundreds of thousands of transistors were placed on a single chip, became more and more common.

Many companies, some new to the computer field, introduced programmable minicomputers supplied with software packages in the 1970s. The "shrinking" trend continued with the introduction of Personal Computers (PCs), which are programmable machines small enough and inexpensive enough to be purchased and used by individuals.

One significant innovation of the fourth generation is the placement of multiple processors on a single machine. Other significant innovations include communications between terminals and computers, and communications over extended networks.

## *New words & Expressions:*

    automatic    自动的
    mechanical    机械的
    difference engine    差分机

program-controlled 程控
analytical engine 分析机
prototype 原型
vacuum tube 真空管
bulky 笨重的
transistor 晶体管
magnetic core 磁芯
product line 产品线
integrated circuits 集成电路
semiconductor memory 半导体存储器
microprocessor 微处理器
silicon chip 硅片
Very Large Scale Integration(VLSI) 超大规模集成电路
terminal 终端

## Exercises

1. Answer the following questions according to the text.

(1) How does the first mechanical calculator develop and how it works?

(2) What causes the appearance of the Electronic Computer?

(3) Please generally discuss about how the computer evolved from mechanical devices to electronic digital devices.

2. Translate the following sentences into English/Chinese.

(1) 使用真空管和再生电容器存储器是第一代计算机的特点。

(2) 第四代计算机的一项重大创新是在一台计算机上安装了多个处理器。

(3) IBM dominated the market of the second generation. Two of IBM's product lines were especially successful: the large 7000-series, and the small 1400-series.

(4) Introduction of the microprocessors (thousands of integrated circuits placed onto a silicon chip) was the hallmark of the fourth generation computers.

3. Match the items in Column A with the translated versions in Column B.

| A | B |
|---|---|
| (1) algorithmic language | (　) a. 计算机接口技术 |
| (2) basis of software technique | (　) b. 数字信号处理 |
| (3) communication fundamentals | (　) c. 软件工程 |
| (4) computer interface technology | (　) d. 计算机系统结构 |
| (5) computer architecture | (　) e. 软件技术基础 |
| (6) digital signal processing | (　) f. 算法语言 |
| (7) experiment of microcomputer | (　) g. 微机控制技术 |
| (8) digital image processing | (　) h. 通信原理 |
| (9) microcomputer control technology | (　) i. 数字图像处理 |
| (10) software engineering | (　) j. 微机实验 |

## ☞ Section 3　Reading Material：

# Classification of Computer Systems

Any classification of computer systems is an ephemeral thing. It will be at least a year from the time this lesson was first typed to the time you are reading it in class. In that time, some of the physical classifications that we have devised will be outdated. However, we will review some of the classifications that are commonly used and try to incorporate them into an overall scheme. One trend that seems to have been followed since the very early days of data and information processing is that the cost of the largest computing systems has stayed relatively constant. These systems grew in complexity and power, but not significantly in cost.

**Mainframes**

The term most commonly used for the largest general purpose computing system is a mainframe computer. The cost ranges from several hundred thousand to several million dollars.

This cost has stayed relatively constant since the early 1950s. However, the power of the mainframe has increased dramatically.

Mainframe systems are designed for large-scale scientific and commercial applications. Their scientific applications range from long-range weather forecasting to the analysis of complex data from high energy nuclear physics experiments. Typical commercial applications are the very complex airline reservation systems and massive banking systems.

Mainframe systems have very fast processor times and extremely large memories. Usually the CPU is comprised of many special purpose logic units that permit the mainframe to handle many tasks concurrently. How this is accomplished will be explained later. It is not unusual to have several hundred people using a single mainframe at once. In addition to the size of memory and speed of processing, these systems have access to vast amounts of secondary storage. Among the best known of the mainframe manufacturers are IBM, UNISYS, Honeywell, and Control Data Corporation (CDC). Burroughs and Sperry merged in 1986 to form UNISYS. Digital Equipment Corporation (DEC) has recently begun to compete directly in this market.

Mainframes are found in almost all large organizations. Although many applications that were once executed exclusively on these systems have migrated to smaller systems, there are still many applications that require the large capacity which only a mainframe can offer. A central system also helps enforce standards and allows management better control of the corporate information resource.

**Minicomputers**

In the 1960s, as the cost of computing continued to decrease, another type of computing system appeared. Its cost is typically from about thirty thousand dollars to several hundred thousand dollars. These systems are usually called minicomputers. Although the term may be misleading today, a mini often is found doing the work of a mainframe.

Originally, the mini had very limited memory and a single logic unit. The most attractive thing about it was its cost. Relatively inexpensive computing capacity could be purchased with a mini, so management was willing to buy these systems for specialized purposes such as process control. Today many of the capacity of the minis found in industry rivals the mainframes of a few years ago. Many of the applications that were exclusively run on the mainframes are being done on minis and the distinction between the two systems, mainframe and mini. The only distinction that seems to remain is the relative cost. Among the best-known minicomputer manufacturers are Digital Equipment Corporation, Hewlett Packard, SUN and Silicon Graphics.

**Microcomputers**

In the 1970s, one of the most significant things that occurred in the evolution of computing systems was the development of "a processor on a chip". Scientists developed the techniques that allowed all of the functions of the ALU to be placed on a single wafer of silicon and the size is a fingernail. This led to the introduction of the third general category of computing systems called the microcomputer, or personal computer. Early micros had very limited processing capabilities, but as has happened with all other computer developments, they have rapidly grown in power and decreased in cost. A typical system will cost from under one hundred dollars to approximately thousands of dollars.

Microcomputers are widely used as personal workstations in almost all aspects of business, industry, and government. They are found extensively in the home and educational institutions. Although there are microcomputer systems that are capable of serving several users at once, most are used by one person. This has led to the term Personal Computer (PC), which is used interchangeably with the term micro. Among the most popular micro computers are those produced by IBM, Apple, and Tandy-Radio Shack. An entire industry has developed manufacturing PCs that function identically to the IBM PC. These machines are often called PC clones. In 1987, IBM introduced its next generation of PCs, the Personal System 2. It is a family of products that significantly increases the power of the earlier machines.

The typical personal computer has a typewriter like keyboard for input, a television type screen, a separate printer for output, and a hard disk for secondary storage. Larger systems add more secondary storage with larger disk drives or re-writable CDs. Many of the systems have the ability to communicate with other personal computers or minis and mainframes. This makes them extremely versatile tools. The applications available on micros are as varied as on the larger systems; the only restrictions are due to their relatively slow speed and memory size.

# Unit 1  An Overview of Computers

**Supercomputers**

One additional category of computing systems we have not discussed is commonly referred to as the supercomputer. These systems have extremely large memories and fast speed. They cost in the range of millions of dollars. They are most commonly used for very large computational problems. These systems are so complex that they usually have a mainframe computer which functions as the input and output device for the supercomputer. A good definition of a supercomputer is simply the largest and most powerful computer that is available at the present time. Supercomputers have found application in long-range weather forecasting and other very complex problems that require enormous amounts of computation. Manufacturers of the largest supercomputers are Cray Research and Control Data Corporation.

## *New words & Expressions:*

supercomputer    超级计算机
mainframe computer    大型计算机
minicomputer    小型机
microcomputer    微型电脑

## *Exercises*

1. Answer the following questions according to the text.

(1) When did the Microcomputers Age begin?

(2) Most of the supercomputers have the capacities to host multiple operating systems and operate as a number of virtual machines and can thus substitute for several small servers, do they?

(3) Please list the different types of computers.

2. Translate the following sentences into Chinese.

(1) The term most commonly used for the largest general purpose computing system is a mainframe computer.

(2) Scientists developed the techniques that allowed all of the functions of the ALU to be placed on a single wafer of silicon and the size is a fingernail.

## ☞ Section 4  Extended Reading:

# What are the Benefits of Computers in School?

Technology gets a bad rap these days. Teens are spending more time online than ever and becoming depressed in the process. Young children learn to type before they can write, a skill that is said to be essential for brain development. They're all shamed for being addicted to devices, and tech companies are blamed for creating addictive tools. In this dystopian picture, it's easy to take for granted the importance of computers in society and to forget the good that technology brings people today. The benefits of a computer and other tech gadgets in teaching and learning are numerous.

**Advantages of a computer in teaching and learning**

The advantages of a computer in teaching and learning are many. Various online collaboration platforms allow students and teachers to collaborate on projects online, inside and outside the classroom on a computer in school, as well as ask questions, share ideas and continue class discussions beyond the classroom.

Easy access to the internet means instant access to information, allowing students to conduct independent research right in the classroom.

Lecture capturing tools allow instructors to capture their lectures on video for students to review later. Similarly, note taking software allows students to easily take, store and access lecture notes.

Course management platforms allow teachers to organize course resources and students to access their grades online. A shared online class calendar helps students be better organized.

Presentation softwares such as Powerpoint and Keynote allow teachers to present lecture material in an interactive way, as well as allowing students to create their own presentations.

Technology has greatly enhanced testing and student self-assessment. Using technology can greatly enhance students' learning about the concepts of math and science.

With technology, students can visit any location on earth without leaving the classroom, and using software like Google Earth.

Technology allows students to become content creators themselves. They can create wiki pages collaboratively, write student blogs, create videos, webpages and other online content.

**Advantages of networking**

Technology provides numerous ways to network within the class when one-on-one interaction is not possible or practical, as well as connecting with schools and experts around the world.

For example, a teacher can let students use Twitter hashtags to ask questions, which would encourage the participation of shy students who otherwise would be too self-conscious to raise their hand and speak up.

Using a live video service, a class can connect with another school halfway across the globe, chat with a famous astronaut, scientist or another celebrity, or bring an expert as a guest lecturer into your classroom.

## ☞ Section 5  Extended Reading:

# What are the Advantages of Computers in Business?

Computers have tremendously improved the way businesses operate in their respective industries. Technology has advanced so remarkably that those who are not using computers in their business are at a major disadvantage against their competitors. In particular, there are several important advantages that computers can provide to the small businesses.

**Organization**

Computers allow the application of different types of software that can help businesses keep track of their files, documents, schedules and deadlines. Computers also allow businesses to organize all of their information in a very accessible manner. The ability to store large amounts of data on a computer is convenient and inexpensive, and saves space. A computer's ability to allow a company to organize its files efficiently leads to better time management and productivity.

**Self-sufficiency**

Computers have made staff and companies more self-sufficient by allowing them to do tasks that previously had to be outsourced. For example, a company can now use office software to create their own training material. Desktop publishing software can be used to create marketing materials. Online tax and accounting programs allow companies to prepare their own taxes. This allows the dominant operations of a company to remain in-house and empowers the company to become more independent and less susceptible to errors committed by outside parties.

**Cost-effective**

Emerging technology makes new tools and services more affordable, and allows companies to save on their staff payroll and office equipment. Because computers allow work to be done faster and more efficiently, it is possible for a company to hire fewer staff. In addition, with networked and relatively inexpensive computers, companies can store data more easily, save the cost of outside file storage, and can avoid having to purchase many copiers, fax machines, typewriters, and other such items that were used before computers became popular.

Correspondingly, potentially profitable businesses can be started with a smaller overhead cost. Email capabilities decrease postage costs; software applications reduce the need for large accounting departments, while videoconferencing reduces the need for travel. All resources saved will trickle down to the consumers, who are then provided with much more affordable products and service.

**Speed**

Computers help speed up other business operations. The collecting of consumer feedback, ordering of raw materials, and inspection of products are made quicker through the use of computers, allowing companies to operate much faster and to produce better quality results.

**Cheaper research and development**

Research and development's cost will also decrease with the help of computers. Scientific research can now be done using the Internet and computer software applications designed to develop and produce new products and services. For example, instead of a company having to do in-person focus groups on a potential new product or to determine their target market, the company can conduct a widespread online survey for a far lower cost. In addition, new models of a product can be created online using virtual pictures and drawings instead of having to be hand-drawn. These interactive models created using software programs can help bring the product and its features to life for a far lower cost than creating an actual physical model of the given product.

**Sales**

Computers can help generate higher sales and profits for businesses via a company website. Many businesses now operate online and around the clock to allow customers from around the world to shop for their products and services.

## ☞ Section 6  Extended Reading:

# Commencement Speech at Stanford Given by Steve Jobs

I am honored to be with you today at your commencement from one of the finest universities in the world. I never graduated from college. Truth be told, this is the closest I've ever gotten to a college graduation. Today I want to tell you three stories from my life. That's it. No big deal. Just three stories.

The first story is about connecting the dots.

I dropped out of Reed College after the first 6 months, but then stayed around as a drop-in for another 18 months or so before I really quit. So why did I drop out?

It started before I was born. My biological mother was a young, unwed college graduate student, and she decided to put me up for adoption. She felt very strongly that I should be adopted by college graduates, so everything was all set for me to be adopted at birth by a lawyer and his wife. Except that when I popped out they decided at the last minute that they really wanted a girl. So my parents, who were on a waiting list, got a call in the middle of the night asking, "We have an unexpected baby boy; do you want him?" They said, "Of course." My biological mother later found out that my mother had never graduated from college and that my father had never graduated from high school. She refused to sign the final adoption papers. She only relented a few months later when my parents promised that I would someday go to college.

And 17 years later I did go to college. But I naively chose a college that was almost as expensive as Stanford, and all of my working-class parents' savings were being spent on my college tuition. After six months, I couldn't see the value in it. I had no idea what I wanted to do with my life and no idea how college was going to help me figure it out. And here I was spending all of the money which my parents had saved on their entire life. So I decided to drop out and trust that it would all work out OK. It was pretty scary at the time, but looking back it was one of the best decisions I ever made. The minute I dropped out, I could stop taking the required classes that didn't interest me, and begin dropping in on the ones that looked interesting. It wasn't all romantic. I didn't have a dorm room, so I slept on the floor in friends' rooms, I returned coke bottles for the 5¢ deposits to buy food with, and I would walk the 7 miles across town every Sunday night to get one good meal a week at the Hare Krishna temple. I loved it. And much of what I stumbled into by following my curiosity and intuition turned out to be priceless later on. Let me give you one example.

Reed College at that time offered perhaps the best calligraphy instruction in the country. Throughout the campus every poster, every label on every drawer, was beautiful hand calligraphed. Because I had dropped out and didn't have to take the normal classes, I decided to take a calligraphy class to learn how to do this. I learned about serif and san serif typefaces, knew about varying the amount of space between different letter combinations, and what makes great typography greater. It was beautiful, historical, artistically subtle in a way that science can't capture, and I found it fascinating.

None of this had even a hope of any practical application in my life. But ten years later, when we were designing the first Macintosh computer, it all came back to me. And we designed it all into the Mac. It was the first computer with beautiful typography. If I had never dropped in on that single course in college, the Mac would have never had multiple typefaces or proportionally spaced fonts. And since Windows just copied the Mac, its likely that no personal computer would have them. If I had never dropped out, I would have never dropped in on this calligraphy class, and personal computers might not have the wonderful typography that they do. Of course, it was impossible to connect the dots looking forward when I was in college. But it was very, very clear looking backwards ten years later.

Again, you can't connect the dots looking forward, you can only connect them looking backwards. So you have to trust that the dots will somehow connect in your future. You have to trust in something — your gut, destiny, life, karma, whatever. This approach has never let me down, and it has made all the difference in my life. My second story is about love and loss.

I was lucky — I found what I loved to do early in life. Woz and I started Apple in my parents garage when I was 20. We worked hard, and in 10 years Apple had grown from just the two of us in a garage into a $2 billion company with over 4000 employees. We had just released our finest creation — the Macintosh — a year earlier, and I had just turned 30. And then I got fired. How can you get fired from a company you started? Well, as Apple grew we hired someone who I thought was very talented to run the company with me, and for the first year or so things went well. But then our visions of the future began to diverge and eventually we had a falling out. When we did, our Board of Directors sided with him. So at 30 years old, I was out, and very publicly out. What had been the focus of my entire adult life was gone, and it was devastating. I really didn't know what to do for a few months. I felt that I had let the previous generation of entrepreneurs down - that I had dropped the baton as it was being passed to me. I met with David Packard and Bob Noyce and tried to apologize for screwing up so badly. I was a very public failure, and I even thought about running away from the valley. But something slowly began to dawn on me — I still loved what I did. The turn of events at Apple had not changed that one bit. I had been rejected, but I was still in love. And so I decided to start over.

I didn't see it then, but it turned out that getting fired from Apple was the best thing that could have ever happened to me. The heaviness of being a successor was replaced by the lightness of being a beginner again, less sure about everything. It freed me to enter one of the most creative periods of my life.

During the next five years, I started a company named NeXT, another company named Pixar, and fell in love with an amazing woman who would become my wife. Pixar went on to create the worlds first computer animated feature film, Toy Story, and is now the most successful animation studio in the world. In a remarkable turn of events, Apple bought NeXT, I returned to Apple, and the technology we developed at NeXT is at the heart of Apple's current renaissance. And Laurene and I have a wonderful family together.

I'm pretty sure none of this would have happened if I hadn't been fired from Apple. It was awful tasting medicine, but I guess the patient needed it. Sometimes life hits you in the head with a brick. Don't lose faith. I'm convinced that the only thing that kept me going on was that I loved what I did. You've got to find what you love. And that is as true for your work as it is for your lovers. Your work is going to fill a large part of your life, and the only way to be truly satisfied is to do what you believe is great work. And the only way to do great work is to love what you do. If you haven't found it yet, keep looking. Don't settle. As with all matters of the heart, you'll know when you find it. And, like any great relationship, it just gets better and better as the years roll on. So keep looking until you find it. Don't settle. My third story is about death.

When I was 17, I read a quote that went something like, "If you live each day as if it was your last, someday you'll most certainly be right." It made an impression on me, and since then, for the past 33 years, I have looked in the mirror every morning and asked myself, "If today were the last day of my life, would I want to do what I am about to do today?" And whenever the answer has been "No" for too many days in a row, I know I need to change something.

Remembering that I'll be dead soon is the most important tool I've ever encountered to help me make the big choices in life. Because almost everything — all external expectations, all pride, all fear of embarrassment or failure — these things just fall away in the face of death, leaving only what is truly important. Remembering that you are going to die is the best way I know to avoid the trap of thinking you have something to lose. You are already naked. There is no reason not to follow your heart.

About a year ago I was diagnosed with cancer. I had a scan at 7:30 in the morning, and it clearly showed a tumor on my pancreas. I didn't even know what a pancreas was. The doctors told me this was almost certainly a type of cancer that was incurable, and that I should expect to live no longer than three to six months. My doctor advised me to go home and get my affairs in order, which was doctor's code for preparing to die. It means to try to tell your kids everything you thought you'd have the next 10 years to tell them in just a few months. It means to make sure everything is buttoned up so that it will be as easy as possible for your family. It means to say your goodbyes.

I lived with that diagnosis all day. Later that evening I had a biopsy, where they stuck an endoscope down my throat, through my stomach and into my intestines, and put a needle into my pancreas and got a few cells from the tumor. I was sedated, but my wife, who was there, told me that when they viewed the cells under a microscope, the doctors started crying because it turned out to be a very rare form of pancreatic cancer that was curable with surgery. I had the surgery and I'm fine now.

This was the closest I've been faced to death, and I hope it's the closest I have gotten for a few more decades. Having lived through it, now I can say this to you with a bit more certainty than when death was a useful but purely intellectual concept: No one wants to die. Even people who want to go to heaven don't want to die to get there. And yet death is the destination we all share. No one has ever escaped it. And that is as it should be, because death is very likely to be the single best invention of life. It is life's change agent. It clears out the old to make way for the new. Right now the new is you, but someday not too long from now, you will gradually become the old and be cleared away. Sorry to be so dramatic, but it is quite true.

Your time is limited, so don't waste it living someone else's life. Don't be trapped by dogma — which is living with the results of other people's thinking. Don't let the noise of others' opinions drown out your own inner voice. And most important, have the courage to follow your heart and intuition. They somehow already know what you truly want to become. Everything else is secondary.

When I was young, there was an amazing publication called The Whole Earth Catalog, which was one of the bibles of my generation. It was created by a fellow named Stewart Brand not far from here in Menlo Park, and he brought it to life with his poetic touch. This was in the late 1960's, before personal computers and desktop publishing, so it was all made with typewriters, scissors, and polaroid cameras. It was sort of like Google in paperback form, 35 years before Google came along: it was idealistic, and overflowing with neat tools and great notions.

Stewart and his team put out several issues of The Whole Earth Catalog, and when it had run its course, they put out a final issue. It was the mid-1970s, and I was your age. On the back cover of their final issue was a photograph of an early morning country road (you might find yourself hitchhiking on if you were so adventurous), beneath it were the words, "Stay Hungry. Stay Foolish." It was their farewell message as they signed off. "Stay Hungry. Stay Foolish." And I have always wished that for myself. And now, as you graduate to begin anew, I wish that for you. Stay Hungry. Stay Foolish.

Thank you all very much.

# Unit 2

## Computer Hardware

After reading this unit and completing the exercises, you will be able to
- Be familiar with the composition of a personal computer.
- Identify the basic hardware components of a desktop personal computer.
- Apply your knowledge when using with a computer for your job.

### ☞ Section 1  Situational Dialogue:

#### Teens and Computers

Todd: OK, Jeanna, you like the computer!

Jeanna: Yes, I do.

Todd: OK. Talk to us about computers.

Jeanna: Well, I go on the computer a lot and I talk with friends through AOL instant messenger. And I just moved from my hometown to Sacramento, so it's a good way to keep in touch with old friends.

Todd: Yeah. Do you learn about computers at school or on your own?

Jeanna: I picked most of what I know, I've picked it up, through, ya know, the years, and some at school, like keyboarding and such.

Todd: OK. Do you have a laptop or a PC?

Jeanna: I have a PC. A Compac.

Todd: Do you like your computer or do you want a new one?

Jeanna: I want a new one cause I think I screwed mine up and it's a little bit slow now.

Todd: OK. How long have you had your computer?

Jeanna: I've had it for about 3 or 4 years.

Todd: Oh, yeah, that's pretty old for a computer. Yeah, so do you talk to your friends?

Jeanna: Yeah, yes I do.

Todd: OK.

Jeanna: See going on and you know the gossip.

Todd: So nowadays, do high school kids talk by e-mail more than phone?

Jeanna: Most people talk by either e-mail or cellular phone. You know a lot of people don't use their house phones as much, and a lot of people have cellular phones.

Todd: OK. Thanks a lot, Jeanna.

## *New words & Expressions:*

AOL instant messenger    AOL 即时通信
laptop    笔记本电脑
gossip    八卦
cellular phone    手机

## ☞ Section 2    Reading Material：

# The Hardware

    Hardware can be categorized in numerous ways. Below we will look at the central unit, or central processing unit (CPU), and the associated equipment consisting of auxiliary storage, input and output devices. All of the units that are connected to the central processing unit are called peripheral devices. In this section we will consider each component in turn.

**CPU**

The part of the computer system that performs the logic transformations on the data and controls all the other devices is the central processing unit, or CPU. The CPU interprets and executes the program instructions. It has three main components: the arithmetic-logic unit (ALU), the control unit, and the primary storage unit. In many personal computers, the ALU and control unit are fabricated as a single component on a single silicon chip with the primary storage on separate chips. For this reason, some people refer to the CPU as just the ALU and control unit and treat primary memory as a separate device.

The control unit serves to coordinate and control the entire operation of the computer system. It controls the transfer of data between the CPU, primary storage, and possibly other auxiliary devices that will be discussed later. The control unit decodes instructions and keeps track of which instruction is to be executed next. In a modern computing system, there may be hundreds or thousands of hardware components. It is the function of the control unit to coordinate their activities. A VDT (video display terminal) is usually employed by the user to monitor the computer's functions and to send control commands to the CPU. With large systems, this person would be a regular employee called the operator working at a special VDT called a console. The control unit examines each instruction and then routes the proper numbers to the arithmetic-logic unit (ALU). The arithmetic-logic unit consists of the circuitry that performs the specified logical or arithmetic operations.

**Memory**

The primary storage unit, or main memory, stores all data and instructions that are data and instructions before they can to be used by the CPU. Main memory is divided into a large number of storage locations used by the CPU. A fixed size of storage locations is called word. Each word can be accessed by the control unit because it has a unique address. The word's address is just like your postal address except it is a single unique number. Each word is made up of a fixed number of units called bytes. In most systems, a byte can hold one character of data. Often the hardware logic permits access to individual bytes and it appears to the user that the computer is "byte addressable," when, in fact, a full word has been accessed in memory.

The size of a computer's memory is usually expressed in terms of bytes. The metric abbreviations for one thousand one million and one billion, K, M and G, are used in specifying memory size. These actually stand for kilobyte, megabyte and gigabyte respectively. However, because of the binary nature of today's computers, it is usual to have K as $2^{10}$ (or 1 024), M as $2^{20}$ (or 1 048 576) and G as $2^{30}$ (or 1 073 741 824). Therefore, a 128M computer has approximately 128 000 000 bytes, or characters, of memory. The size of the memory is one of the parameters that determines the power and cost of the computer.

**Auxiliary storage devices**

In today's modern computing systems, it is necessary to have another form of memory. It is often the case that the main memory is not large enough to accommodate all the data necessary for a particular application. It is also the case that main memory is only used to hold data that are currently being processed. When not in use, data are stored on auxiliary storage devices.

Auxiliary storage devices (sometimes called secondary storage) can overcome this difficulty. The most common devices used are magnetic disk drives, CDs or magnetic tapes used for backups. Many personal computers still use a flexible disk called a floppy. Common memory sizes for personal computers are in the order of several tenths of M, while associated hard disk drives may contain tenths of gigabytes of storage.

Auxiliary storage devices can store very large quantities of data in an economical and reliable manner. Their major limitation is that the data are not available to the CPU nearly as quickly as are data stored in main memory, and care must be taken to balance the use of main and auxiliary memory. Main memory may be up to 100 000 faster than auxiliary memory.

**Input/Output devices**

Information we normally deal with in our daily lives, such as printed material we read or voice information we hear on the telephone, is essentially useless to the computer in the form that we understand. Devices that convert data into a form that the computer can interpret (computer-readable) and enter it into the computer are called input devices. There is a wide variety of input devices.

Among the ones we will consider are keyboard devices, scanning devices, and audio or voice recognition devices. As in other areas of computing, technological advances in input devices continue to improve accuracy and speed while reducing costs.

Output devices serve two important functions:

- to create new computer-readable media that can be utilized in subsequent data and information processing steps and
- to transform the internal information into a form that we can understand, such as printed reports or voice messages.

Some familiar output devices include printers and plotters.

## *New words & Expressions:*

component  组件，元件；部分
motherboard  主板
CPU (Central Processing Unit)  中央处理器
memory  内存
storage device  存储设备
input device  输入设备
output device  输出设备
peripheral  外围的；外围设备
capacity  容量
command  命令，指令
instruction  指令
symbolic  符号的，象征的
ALU (Arithmetic Logic Unit)  算术逻辑单元
mechanism  机制，原理

## *Exercises*

1. Answer the following questions according to the text.

(1) What kinds of components do you need to assemble a desktop computer?

(2) What's the use of a computer memory?

(3) What does "CPU" stand for?

(4) What's the use of input devices? What's the use of output devices? What is the distinctness between in both?

2. Translate the following sentences into English/Chinese.

(1) 主存储器分为大量的可被 CPU 使用的存储单元。

(2) Devices that convert data into a form that the computer can interpret (computer-readable) and enter it into the computer are called input devices. There is a wide variety of input devices. Among the ones we will consider are keyboard devices, scanning devices, and audio or voice recognition devices.

3. Match the items in Column A with the translated versions in Column B.

| A | B |
|---|---|
| (1) CPU | ( ) a. 扩展卡 |
| (2) power supply | ( ) b. 内存 |
| (3) I/O ports | ( ) c. 中央处理器 |
| (4) CD-ROM | ( ) d. 文件配置表 |
| (5) FAT | ( ) e. 硬盘 |
| (6) memory | ( ) f. 基本输入输出系统 |
| (7) hard disk | ( ) g. 输入/输出端口 |
| (8) BIOS | ( ) h. 只读光盘 |
| (9) motherboard | ( ) i. 主板 |
| (10) expansion card | ( ) j. 电源 |

## ☞ Section 3  Reading Material:

# The Motherboard

The Motherboard serves to connect all of the parts of a computer together. The CPU, memory, hard drives, and other ports and expansion cards all connect to the motherboard directly or via cables.

The motherboard is the piece of computer hardware that can be thought of as the "backbone" of the PC, or more appropriately as the "mother" that holds all the pieces together.

Phones, tablets and other small devices have motherboards too but they're often called logic boards instead. Their components are usually soldered directly onto the board to save space, which means there aren't expansion slots for upgrades like you see in desktop computers.

The IBM Personal Computer that was released in 1981 is considered to be the very first computer motherboard (it was called a "planar" at the time).

Popular motherboard manufacturers include ASUS, AOpen, Intel, ABIT, MSI, Gigabyte, and Biostar.

**Motherboard components**

Everything behind the computer case is connected in some way to the motherboard so that all the pieces can communicate with each other.

This includes video cards, sound cards, hard drives, optical drives, the CPU, RAM sticks, USB ports, a power supply, etc. On the motherboard are also expansion slots, jumpers, capacitors, device power and data connections, fans, heat sinks, and screw holes.

**Important motherboard facts**

Desktop motherboards, cases and power supplies all come in different sizes called form factors. All three must be compatible to work properly together.

Motherboards vary greatly with respect to the types of components they support. For example, each motherboard supports a single type of CPU and a short list of memory types. Additionally, some video cards, hard drives, and other peripherals may not be compatible. The motherboard manufacturer should provide clear guidance on the compatibility of components.

In laptops and tablets, and increasingly even in desktops, the motherboard often incorporates the functions of the video card and sound card. This helps keep these types of computers small in size. However, it also prevents those built-in components from being upgraded.

Poor cooling mechanisms in place for the motherboard can damage the hardware attached to it. This is why high-performance devices like the CPU and high-end video cards are usually cooled with heat sinks, and integrated sensors are often used to detect the temperature and communicate with the BIOS or operating system to regular the fan speed.

Devices connected to a motherboard often need device drivers manually installed in order to make them work with the operating system. See How to Update Drivers in Windows if you need help.

**Physical description of a motherboard**

In a desktop, the motherboard is mounted inside the case, opposite the most easily accessible side. It is securely attached via small screws through pre-drilled holes.

The front of the motherboard contains ports that all of the internal components connect to. A single socket/slot houses the CPU. Multiple slots allow for one or more memory modules to be attached. Other ports reside on the motherboard, and these allow the hard drive and optical drive (and floppy drive if present) to connect via data cables.

Small wires from the front of the computer case connect to the motherboard to allow the power, reset, and LED lights to function. Power from the power supply is delivered to the motherboard by use of a specially designed port.

Also on the front of the motherboard are a number of peripheral card slots. These slots are where most video cards, sound cards, and other expansion cards are connected to the motherboard.

On the left side of the motherboard (the side that faces the back end of the desktop case) are a number of ports. These ports allow most of the computer's external peripherals to connect such as the monitor, keyboard, mouse, speakers, network cable and more.

All modern motherboards also include USB ports, and increasingly other ports like HDMI and FireWire, that allow compatible devices to connect to your computer when you need them—devices like digital cameras, printers, etc.

The desktop motherboard and case are designed so that when peripheral cards are used, the sides of the cards fit just outside the back end, making their ports available for use.

## *New words & Expressions:*

expansion card　　扩展卡

## *Exercises*

1. Answer the following questions according to the text.

(1) Which is heart of the computer, with all of its connections leading out from itself and into every device in the machine?

(2) What are motherboard components?

(3) Please list the common use of the motherboards.

2. Translate the following sentences into English/Chinese.

(1) 主板将计算机的各个部分联系到一起。

(2) The motherboard is the piece of computer hardware that can be thought of as the "backbone" of the PC, or more appropriately as the "mother" that holds all the pieces together.

3. Match the items in Column A with the translated versions in Column B.

A　　　　　　　　　　　B

a. 打印机　　　　　　　(　) (1) mouse

b. 主板　　　　　　　　(　) (2) printer

c. 显示器　　　　　　　(　) (3) computer case

d. 鼠标　　　　　　　　(　) (4) monitor

e. 主机箱　　　　　　　(　) (5) motherboard

## ☞ Section 4　Extended Reading:

# How Bluetooth is Making Our Lives Easier

When you think of Bluetooth technology, you may be thinking of the Bluetooth headsets used to take calls when you are driving or simply don't want to have the phone against your ear. Many people are now seeing Bluetooth in a whole different light and are now using it to make things easier to do than ever before.

Bluetooth has broadened its horizons and found itself almost indispensable in many different areas of technology today.

Here are five examples of how Bluetooth technology is making our lives easier.

**Streaming music to your sound system**

There is something incredibly convenient about being able to relax in your bed, streaming music from your mobile to your sound system on the other side of the room. With the possibilities that services, such as Spotify, offer us in producing one-of-a-kind playlists, it is now easier than ever to enjoy that music without compromise, and Bluetooth just adds to the experience, meaning we no longer have to get out of bed to change the CD.

There are even external Bluetooth devices available to revamp old traditional speakers, giving them the ability to accept Bluetooth signals like modern Bluetooth-enabled speakers.

**Tracking your every step**

If you want to be able to track your child's every step, Bluetooth stickers mean that you can see their every movement in real time on your smart phone. Although this is not for everyone, it can prove very useful for when you have accidentally lost track of your child in a busy place or have a missing pet that may just be in your next-door neighbor's garden.

**Opening doors**

Have you ever closed the front door behind yourself, just moments later to realize you left the keys on the kitchen table and won't be able to get back in after work? If you are forgetful with your keys, then you may find it interesting to know that you can use a Bluetooth-driven smart phone app to unlock doors without the need for your keys. This is pretty useful if you can't afford to get a new lock fitted and a set of new keys cut. You can also use the app to open your garage door if it is raining and you don't want to leave the car.

**Wireless gaming**

Tethered gaming handsets were restricting and meant that you had a limit of how far you could sit from the gaming console. PS3 and Nintendo Wii are strong advocates of wireless gaming, allowing their users to have a more realistic gaming experience. As televisions have got bigger, it has become more important to sit further away from the screen, without pulling the games console off the shelf.

**Driving music**

Just as we all love the convenience of being able to stream music from our bed to our bedroom sound system, the same principle applies to car sound systems. When you are driving and want more choice than your collection of 3 CDs from years ago or the radio, it can be very convenient to stream all of your favorite music from your smart phone to your car's sound system.

I have only covered five ways in which Bluetooth technology is making our everyday lives easier, but there are many more ways. One of the main areas we are seeing Bluetooth being used to great effect is the health and fitness industry, with a wide range of gadgets to monitor our activity. With the technology surrounding Bluetooth modules continuously evolving, who knows what Bluetooth will offer us in ten years' time.

☞ Section 5  Extended Reading:

# Alibaba Group will Invest $15B into a New Global Research and Development Program

Alibaba Group announced today that it plans to invest more than $15 billion over the next three years into a global research and development initiative called Alibaba DAMO Academy. The Chinese tech giant said the program, which is currently recruiting 100 researchers, will help it reach its goal of serving two billion customers and creating 100 million jobs by 2036, while also "increasing technological collaboration worldwide."

DAMO Academy (the initials stand for "discovery, adventure, momentum and outlook") will be led by Alibaba Group chief technology officer Jeff Zhang and start by opening labs in seven cities around the world: Beijing and Hangzhou in China; San Mateo and Bellevue in the U.S.; Moscow, Russia; Tel Aviv, Israel; and Singapore.

Alibaba's researchers will collaborate closely with university programs such as U.C. Berkeley's RISE Lab, which is developing technologies that enable computers to make secure decisions based on real-time data. DAMO Academy's current advisory board also includes professors from Princeton, Harvard, MIT, the University of Washington, Columbia University, Beijing Institute of Technology, Peking University and Zhejiang University.

Research will focus on a wide array of areas, including data intelligence, the Internet of Things, financial tech, quantum computing and human-machine interaction. More specifically, Alibaba Group said researchers will look at machine learning, network security, visual computing and natural language processing.

Alibaba joins other major Chinese tech firms in setting up labs and working closely with technology researchers and universities in the U.S. and other countries. These programs give companies the benefit of dipping into new talent pools without having to convince potential hires to move to China (and also potentially luring them away from rivals like Amazon, Facebook, Google and Apple).

For example, Baidu's research lab in Silicon Valley, which works on big data, deep learning and artificial intelligence, is recruiting from top American research universities, while Tencent (the maker of WeChat) announced plans in May to set up an AI research lab in Seattle. Huawei also set up an AI research partnership with U.C. Berkeley last year with initial funding of $1 million.

## ☞ Section 6  Extended Reading:

# 10 Ways to Make a Computer Move Faster

The longer a computer is in use, the slower it gets. Even if you have taken excellent care of it, you can not entirely safeguard it against slowing down. However, there are quite a few ways to improve its quality of life by making it move faster.

**Clean out some junk**

You can clear out unnecessary data stored in your computer by removing the cache and cookies in your browser. If you use Internet Explorer, you will go to Tools, then click on the Internet Options, then choose to delete Browser History. In Mozilla Firefox, click on Tools and then click on Clear Recent History.

**Defragment**

Use causes hard drive data to scatter. This disk fragmentation will cause your computer to become slower or even crash. Run the Windows Disk Defragmenter utility to put all the pieces back together on your hard drive and make file access more efficient. In the Start menu, choose "Run...", and type "defrag" to access the applet.

**Virus scan**

Although sometimes a virus does not completely destroy your computer, it can make the computer slow. Please install an antivirus program and run scans frequently.

**Adware/Spyware scan**

Malicious software may have crept into your computer without your knowledge. Install anti-adware and anti-spyware programs, and run frequent scans.

**Boost your startup**

Your computer may be running unnecessary programs when it starts up, causing a bottleneck when you boot up. Type "msconfig" in the "Start/Run..." dialog box click on the Startup tab. Uncheck the boxes of programs you do not use frequently. Avoid disabling a program if you aren't sure what it is, as it could be a vital system process.

**Update Windows**

By installing new security patches and updates from Windows, you are doing your part to ensure that your computer and drivers run more efficiently and securely. Some computers come with updates that pop up on occasion and all you have to do is click "OK" to keep your computer up to date. However, you may also have to visit the Windows website online to get needed updates.

**Close unused programs**

Close programs you are not using. Open programs take up processor and RAM resources even if they aren't actively being used.

**Remove unnecessary stuff**

You can free up disk space by uninstalling some programs. Seriously consider the perks of having something you have not used in six months as opposed to having space for other things you would like to run regularly. If you are running Windows XP, please go to Control Panel and click on "Add/Remove Programs". If you are running Windows Vista, go to "Control Panel" and click on "Programs and Features".

**Add memory**

Having a sufficient store of memory allows your computer to run faster (while your hard drive is where your files are held). If you right click on the "My Computer" icon, left click on "Properties", and your memory (RAM) will show up at the bottom of the window that you see. Having at least 1 GB of memory should suit your needs, but you may need 2 GB if, you often use a video editing program.

**Restart**

Restart your computer weekly to address unresolved performance issues. It can help free up some memory and could be the solution to many problems.

# Unit 3

## Operating System

After reading this unit and completing the exercises, you will be able to
- Understand the basic function of operating system.
- Be familiar with the different types of operating system.
- Identify some operating systems, such as Windows 10, Linux, and so on.
- Apply your knowledge when using your operating system.

### ☞ Section 1  Situational Dialogue:

## System Crash

Cheng Hong has met with some problems with his computer, and now he is consulting Qian Liang, a "geek" among his classmates.

Cheng Hong: Good morning, Qian Liang. May I ask you some questions about my computer?

Qian Liang: Good morning. What can I do for you?

Cheng Hong: The system in my computer has crashed. I have tried several times to reboot it, but it doesn't work. What can I do?

Qian Liang: How did that happen?

Cheng Hong: The screen went blank while I was just drawing with Photoshop. It all happened suddenly, and I am totally at a loss why this happened.

Qian Liang: Well, there might be many causes for system crash. One of them is software. The simplest operation like installing and uninstalling might cause a crash in system. Problems within the system may also do that, so you have to be very careful when deleting DLL files, modifying Windows registry, or upgrading the operating system. Another cause is the hardware problem. Have you tried backing up the files? It would be a great help in system restoration when the problem doesn't lie in the hardware.

Cheng Hong: The service personnel installed a ghost program for me when I bought the computer. Will I lose the data in my hard disk if it is restored?

Qian Liang: Not except the data on C: drive, where files on your desktop or in My Documents are stored. You may move them into other locations with a bootable disk or under DOS if there are any.

Cheng Hong: There are a lot of learning materials in My Documents and on the desktop. Would you please help me transfer them? It might save me the trouble of calling for customer services, and what's more important, I'd learn a lot from you.

Qian Liang: No problem.

## *New words & Expressions:*

consult    咨询
geek    极客
crash    崩溃
reboot    重启
install    安装
uninstall    卸载
modifying    修改
registry    注册表
upgrade    更新
back up    备份
restoration    恢复

## ☞ Section 2  Reading Material:

# Operating System

An operating system (sometimes abbreviated as "OS") is a software program that enables the computer hardware to communicate and operate with the computer software.

It controls and coordinates the use of the hardware among the various applications programs for various uses. It acts as a resource allocator and manager, like Figure 3-1.

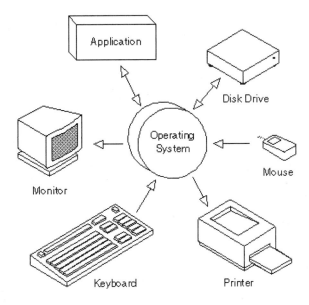

Figure 3-1

Types of operating systems are shown in the following texts.

**Single-tasking and multi-tasking**

A single-tasking operating system can only run one program at a time, while a multi-tasking operating system allows more than one program to be running in concurrency. This is achieved by time-sharing, where the available processor time is divided between multiple processes. These processes are each interrupted repeatedly in time slices by a task-scheduling subsystem of the operating system. Multi-tasking may be characterized in preemptive and co-operative types. In preemptive multi-tasking, the operating system slices the CPU time and dedicates a slot to each of the programs. Unix-like operating systems, such as Solaris and Linux—as well as non-Unix-like, such as AmigaOS—support preemptive multi-tasking. Cooperative multi-tasking is achieved by relying on each process to provide time to the other processes in a defined manner. 16-bit versions of Microsoft Windows used cooperative multi-tasking. 32-bit versions of both Windows NT and Win9x, used preemptive multi-tasking.

**Single-user and multi-user**

Single-user operating systems have no facilities to distinguish users, but may allow multiple programs to run in tandem. A multi-user operating system extends the basic concept of multi-tasking with facilities that identify processes and resources, such as disk space, belonging to multiple users, and the system permits multiple users to interact with the system at the same time. Time-sharing operating systems schedule tasks for efficient use of the system and may also include accounting software for cost allocation of processor time, mass storage, printing, and other resources to multiple users.

**Distributed**

A distributed operating system manages a group of distinct computers, and makes them appear to be a single computer. The development of networked computers that could be linked and communicate with each other gave rise to distributed computing. Distributed computations are carried out on more than one machine. When computers in a group work in cooperation, they form a distributed system.

**Templated**

In an OS, distributed and cloud computing context, templating refers to creating a single virtual machine image as a guest operating system, then saving it as a tool for multiple running virtual machines. The technique is used both in virtualization and cloud computing management, and is common in large server warehouses.

### Embedded

Embedded operating systems are designed to be used in embedded computer systems. They are designed to operate on small machines like PDAs with less autonomy. They are able to operate with a limited number of resources. They are very compact and extremely efficient by design. Windows CE and Minix 3 are some examples of embedded operating systems.

### Real-time

A real-time operating system is an operating system that guarantees to process events or data by a specific moment in time. A real-time operating system may bea single-tasking or multi-tasking. When it is a multi-tasking, it uses specialized scheduling algorithms so that a deterministic nature of behavior is achieved. An event-driven system switches between tasks based on their priorities or external events while time-sharing operating systems switch tasks based on the clock interrupts.

### Library

A library operating system is one in which the services that a typical operating system provides, such as networking, are provided in the form of libraries and composed with the application and configuration code to construct a unikernel: a specialized, single address space, machine image that can be deployed to cloud or embedded environments.

There are still more types of OSs according to different classifications, such as Network Operating System,Time Sharing Operating System, Desktop Operating System, Server Operating System, and so on.

The most widely used contemporary desktop operating systems include Microsoft Windows, Macintosh, and Unix-like systems such Linux and the BSD(Berkeley Software Distribution ).

## *New words & Expressions:*

batch processing operating system 成批量操作系统

real-time operating system　　实时操作系统
single user operating system　　单一操作系统
multi-tasking operating system　　多任务操作系统
multi-user operating system　　多用户操作系统
distributed operating system　　分布式操作系统

## Exercises

1. Answer the following questions according to the text.

(1) How do you think about system software and operating system?

(2) What is the most popular operating system in our everyday life?

(3) Please list the common OS.

2. Translate the following sentences into English/Chinese.

(1) 操作系统扮演了资源分配者和管理者的角色。

(2) Single-user operating systems have no facilities to distinguish users, but may allow multiple programs to run in tandem.

(3) An operating system (sometimes abbreviated as "OS") is a software program that enables the computer hardware to communicate and operate with the computer software.

3. Match the items in Column A with the translated versions in Column B.

A　　　　　　　　　　　　　　B

a. 批处理　　　　　　　　　(　　) (1) multi-tasking

b. 多任务　　　　　　　　　(　　) (2) multi-processor

c. 内核　　　　　　　　　　(　　) (3) batch

d. 外壳　　　　　　　　　　(　　) (4) shell

e. 多处理器　　　　　　　　(　　) (5) kernel

4. Choose the right answer

(1) Which one is not an operation system?

A. Mac OS X　　　B. Unix　　　C. API　　　D. Linux

(2) Operating systems are loaded in many electronic devices except _____.

A. cell phones          B. microwaves          C. laptop          D. tablet PC

(3) What cannot an operating system do?

A. It manages the hardware.

B. It manages the software.

C. It distributes the resources.

D. It acts as the CPU of the computers.

(4) The operating system plays the role of the good parent because _____.

A. it is always fair

B. it is inherited by other applications

C. it makes sure that each application gets the necessary resources

D. it accommodates thousands of different printers, disk drives and special peripherals in any possible combination

(5) _____ is charged with managing the hardware and the distribution of its resources.

A. Application          B. API          C. CPU          D. The operating system

## ☞ Section 3  Reading Material:

## An Overview on the Most Commonly-used OS

**Microsoft Windows**

Microsoft Windows is a family of proprietary operating systems designed by Microsoft Corporation and primarily targeted to Intel architecture based computers, with an estimated 88.9 percent total usage share on Web connected computers. The latest version is Windows 10.

In 2011, Windows 7 overtook Windows XP as most common version in use.

Microsoft Windows was first released in 1985, as an operating environment running on top of MS-DOS, which was the standard operating system shipped on most Intel architecture personal computers at the time. In 1995, Windows 95 was released which only used MS-DOS as a bootstrap. For backwards compatibility, Win9x could run real-mode MS-DOS and 16-bit Windows 3.x drivers. Windows ME, released in 2000, was the last version in the Win9x family.

Later versions have all been based on the Windows NT kernel. Current client versions of Windows run on IA-32, x86-64 and 32-bit ARM microprocessors. In addition Itanium is still supported in older server version Windows Server 2008 R2. In the past, Windows NT supported additional architectures.

Server editions of Windows are widely used. In recent years, Microsoft has expended significant capital in an effort to promote the use of Windows as a server operating system. However, Windows' usage on servers is not as widespread as on personal computers as Windows competes against Linux and BSD for server market share.

**Unix and Unix-like operating systems**

Unix was originally written in assembly language. Ken Thompson wrote B, mainly based on BCPL, based on his experience in the MULTICS project. B was replaced by C, and Unix, rewritten in C, developed into a large, complex family of inter-related operating systems which have been influential in every modern operating system.

The Unix-like family is a diverse group of operating systems, with several major sub-categories including System V, BSD, and Linux. The name "UNIX" is a trademark of The Open Group which licenses it for use with any operating system that has been shown to conform to their definitions. "UNIX-like" is commonly used to refer to the large set of operating systems which resemble the original UNIX.

Unix-like systems run on a wide variety of computer architectures. They are used heavily for servers in business, as well as workstations in academic and engineering environments. Free UNIX variants, such as Linux and BSD, are popular in these areas.

Four operating systems are certified by The Open Group (holder of the Unix trademark) as Unix. HP's HP-UX and IBM's AIX are both descendants of the original System V Unix and are designed to run only on their respective vendor's hardware. In contrast, Sun Microsystems's Solaris can run on multiple types of hardware, including x86 and Sparc servers, and PCs. Apple's MacOS, a replacement for Apple's earlier (non-Unix) MacOS, is a hybrid kernel-based BSD variant derived from NeXTSTEP, Mach, and FreeBSD.

Unix interoperability was sought by establishing the POSIX standard. The POSIX standard can be applied to any operating system, although it was originally created for various Unix variants.

## MacOS

MacOS (formerly "Mac OS X" and later "OS X") is a line of open core graphical operating systems developed, marketed, and sold by Apple Inc., the latest of which is pre-loaded on all currently shipping Macintosh computers. MacOS is the successor to the original classic MacOS, which had been Apple's primary operating system since 1984. Unlike its predecessor, MacOS is a UNIX operating system built on technology that had been developed at NeXT through the second half of the 1980s and up until Apple purchased the company in early 1997. The operating system was first released in 1999 as Mac OS X Server 1.0, followed in March 2001 by a client version (Mac OS X v10.0 "Cheetah"). Since then, six more distinct "client" and "server" editions of MacOS have been released, until the two were merged in OS X 10.7 "Lion".

Prior to its merging with MacOS, the server edition — MacOS Server — was architecturally identical to its desktop counterpart and usually ran on Apple's line of Macintosh server hardware. MacOS Server included work group management and administration software tools that provide simplified access to key network services, including a mail transfer agent, a Samba server, an LDAP server, a domain name server, and others. With Mac OS X v10.7 Lion, all server aspects of Mac OS X Server have been integrated into the client version and the product re-branded as "OS X" (dropping "Mac" from the name). The server tools are now offered as an application.

## Linux

The Linux kernel originated in 1991, as a project of Linus Torvalds, while a university student in Finland. He posted information about his project on a newsgroup for computer students and programmers, and received support and assistance from volunteers who succeeded in creating a complete and functional kernel.

Linux is Unix-like, but was developed without any Unix code, unlike BSD and its variants. Because of its open license model, the Linux kernel code is available for study and modification, which resulted in its use on a wide range of computing machinery from supercomputers to smart-watches. Although estimates suggest that Linux is used on only 1.82% of all "desktop" (or laptop) PCs, it has been widely adopted for use in servers and embedded systems such as cell phones. Linux has superseded Unix on many platforms and is used on most supercomputers including the top 385. Many of the same computers are also on Green500 (but in different order), and Linux runs on the top 10. Linux is also commonly used on other small energy-efficient computers, such as smartphones and smartwatches. The Linux kernel is used in some popular distributions, such as Red Hat, Debian, Ubuntu, Linux Mint and Google's Android, Chrome OS, and Chromium OS.

## *New words & Expressions:*

  version  版本
  open-source project 开源项目
  architecture  架构

## *Exercises*

1. Answer the following questions according to the text.

(1) Which was based on the Unix system, with the source code being a part of GNU open-source project?

(2) How many modules does MS/DOS have?

(3) Which operation system is the successor to Windows 2000?

2. Translate the following sentences into English/Chinese.

(1) MacOS (formerly "Mac OS X" and later "OS X") is a line of open core graphical operating systems developed, marketed, and sold by Apple Inc.

(2) Microsoft Windows was first released in 1985, as an operating environment running on top of MS-DOS, which was the standard operating system shipped on most Intel architecture personal computers at the time.

(3) 微软的Windows占领了个人电脑市场上超过90%的市场份额，一举超越在1984年推出的Mac OS操作系统。

3. Match the items in Column A with the translated versions in Column B.

| A | B |
|---|---|
| (1) batch processing OS | (　) a. 服务器操作系统 |
| (2) distributed OS | (　) b. 批处理操作系统 |
| (3) single user OS | (　) c. 开源操作系统 |
| (4) real-time OS | (　) d. 分时操作系统 |
| (5) embedded OS | (　) e. 实时操作系统 |
| (6) multi-tasking OS | (　) f. 多任务操作系统 |
| (7) open source OS | (　) g. 网络操作系统 |
| (8) server OS | (　) h. 单用户操作系统 |
| (9) time sharing OS | (　) i. 分布式操作系统 |
| (10) network OS | (　) j. 嵌入式操作系统 |

## ☞ Section 4  Extended Reading:

# Android Operating System

Android is an operating system based on the Linux kernel with a user interface based on direct manipulation, designed primarily for touchscreen mobiles using touch inputs, that loosely correspond to real-world actions, like swiping, tapping, pinching, and reverse pinching to manipulate on-screen objects, and a virtual keyboard. Despite being primarily designed for touchscreen input, it also has been used in televisions, games consoles, digital cameras, and other electronic devices such as smartphones and tablet computers.

As of 2011, Android has the largest installed base of any mobile OS and as of 2013, its device Android is an operating system based on the Linux kernel with a user interface based on direct manipulation, designed primarily for touchscreen mobile devices such as smartphones and tablet computers, using touch inputs, that loosely correspond to real-world actions, like swiping, tapping, pinching, and reverse pinching to manipulate on-screen objects, and a virtual keyboard.

Despite being primarily designed for touchscreen input, it also has been used in televisions, games consoles, digital cameras, and other electronics. In April — May 2013 it was found that 71% of mobile developers developed for Android.

Android's source code is released by Google under open source licenses, although most Android devices ultimately ship with a combination of open source and proprietary software. Initially developed by Android, Inc., which Google backed financially and later bought in 2005, Android was unveiled in 2007 along with the founding of the Open Handset Alliance—a consortium of hardware, software, and telecommunication companies devoted to advancing open standards for mobile devices.

Android is popular with technology companies which require a ready-made, lowcost and customizable operating system for high-tech devices. Android's open nature has encouraged a large community of developers and enthusiasts to use the open-source code as a foundation for community-driven projects, which add new features for advanced users or bring Android to devices which were officially released running other operating systems. The operating system's success has made it a target for patent litigation as part of the so-called "smartphone wars" between technology companies.

## ☞ Section 5　Extended Reading：

## Comparison between Linux and Windows

1) What is it?

Linux is a computer operating under free and open source software development and distribution model. It was first released on 5 October 1991 by Linus Torvalds.

Whereas Windows operating system is developed by Microsoft and it is most famous OS in the world.

2) Cost

Linux is free of cost, it is freely distributed and it can be downloaded free of cost. Some of the versions of Linux are even charged but the charge is very less as compared to that of

Windows.

Whereas Windows OS for desktops and laptops are generally expensive. A single authorized OS copy may cost $50 to $450, price varies from version to version.

3) Manufacturer

Linux is developed by community and the owner is Linus Torvalds.

Whereas Windows is manufactured by Microsoft but allows many computer manufacturers to sell their computers with Windows preinstalled in them.

4) Users

Linux and Windows both are used by everyone. From home users to developers and computer enthusiasts

5) Supported File system

Linux supports Ext2, Ext3, Ext4, Jfs, ReiserFS, Xfs, Btrfs, FAT, FAT32 and NTFS

Whereas Windows supports FAT, FAT32, NTFS and exFAT

6) GUI provided

Linux provides two GUIs that are KDE and Gnome. But there are many alternatives available like Mate, TWM, Unity etc

Whereas in Windows, the GUI is an integral component of the OS and is not replaceable.

7) Security

Linux is far better in case of security as only about 60 to 70 viruses listed till date in Linux. And if we are talking about Windows, more than 60,000 viruses had been listed till now.

And if we talk about detection and solving the threats than in this case Linux detects and solves it very quickly but Windows takes 2 to 3 months to detect and solve the issue.

8) Processors

Linux supports dozens of processors where as in Windows, limited processors are available.

9) Games

In this case Windows are much better then Linux as very few games are available for Linux whereas millions of games are available for Windows.

10) Ease

Microsoft Windows are easiest of all the operating systems. Microsoft makes several advancements and changes so that it is much easier than any other. Although it is not easiest but it is easier than Linux.

Although Linux is changing dramatically so that it becomes easier but still Windows are far more easier than Linux.

11) Open source

Many of the Linux programs and variants are open source and enables users to modify and customize the code according to their needs.

Whereas in Microsoft Windows are not open source and most of the programs can't be altered as they are not open source.

12) Support

The Linux is very less used hence it is difficult to find out users familiar with Linux and its programs, but users can get vast amount of helps available online on various websites and there are many books available for guiding the user to use Linux.

Where in case of Microsoft Windows, it includes its own helping section and it also has vast amount of helps available online on various websites and there are many books available for guiding the user to use all of the versions of Windows.

Various forms of Linux are Ubuntu, Red Hat, Android, etc.

Whereas in Windows various forms are Windows 8, Windows 8.1, Windows vista, Windows 7, windows XP and many more.

## ☞ Section 6   Extended Reading：

## Should You Upgrade or Repair Your Computer?

Technology develops at an incredible pace. Your shiny new desktop PC could be outdated within 6 months time. Those with a computer over the age of 2 years can fully expect it to start spluttering along as you try and do the simplest of things. However, all hope is not lost. New processors and graphics cards are all well and good, but most computer problems can be fixed by simple updates and a little maintenance. Even if you do need new parts, upgrading a computer has never been easier and so cheap.

**Give it a cleaning**

Never underestimate what a good physical clean can do for a computer. Computers get full of dust and lint, and fans and vents can get clogged. If a computer is suffering from overheating, random shut-downs or noisy whirring, it may not need repairing as the culprit could simply be an abundance of dust. Use short bursts of compressed air to displace dust in hard to reach places and give the rest a wipe with lint-free, anti-static cloths. Of course, unplug the power and be very careful when cleaning delicate components.

**Software repair**

New computers are always much faster than old ones, and it's not just because they have newer components. As a computer is used over time, new pieces of software are installed on it, hard drives get filled and it is running more and more processes each time you boot up. Schedule a regular virus scan to get rid of malware and adware, uninstall programs that are clogging up your desktop and disable rarely used software from booting at the start-up. Alternatively, a complete fresh install to a faulty operating system, while a drastic measure, will really get everything back to running smoothly (make sure you do a full back-up, though). You'd be surprised by how much performance can be gained from repairing an existing system.

**Hardware repair**

If your computer still has performance problems after a software repair, then it could mean the issue is with your computers hardware. Unfortunately hardware can break, with the most common causes being overheating, liquid damage and impact related damage. All is not lost though, as hardware can be easily repaired or replaced by a computer repair technician.

Generally hardware repairs are priced based on the time taken to complete the job and the cost of the replacement part, if one is needed. Along with the expertise needed to successfully make hardware repairs, repair technicians will also be able to easily identify the exact issue with a broken computer.

If the offending item of hardware is easy to identify (e.g. a broken CD drive), then you can always have a go at fixing it yourself; parts can be ordered online through the brand manufacturer.

You can also find second hand parts on eBay, however buying straight from the manufacturer will ensure your new parts have a full warranty. Although fixing it yourself may work out cheaper, you do run the risk of doing further damage to your computer if it's done incorrectly or haphazardly.

**Upgrading**

Unfortunately, there is only so much that uninstalling and repairing can do. If you are still suffering from PC performance issues, a few bits may need upgrading. Fortunately, replacing and upgrading many computer components is very easy, and even technophobes should be able to manage it.

Buying more RAM is perhaps the quickest and most efficient way to boost performance, and it is simply a case of opening up the case and slotting RAM into the motherboard. There are compatibility issues, so check what RAM you currently have before you buy some. If you're a heavy downloader or you find that you keep running out of disk space, a new hard-drive is a practical upgrade, and external hard-drives can simply be plugged into a USB port. For gamers, a new graphics card may be a good upgrade and, while they can be a bit fiddly to fit, practical online help can guide you through the install.

# Unit 4

## Application Software

After reading this unit and completing the exercises, you will be able to
- Understand the different kinds of application software.
- Be familiar with the common application software.
- List the common application software.

## ☞ Section 1  Situational Dialogue:

### A Job Interview

Ms Liu: Excuse me. I have an appointment with Mr. Li at nine. May I come in?

Mr. Li: Yes, come in please. I am Mr. Li. You must be Ms Liu, right?

Ms Liu: Yes, I am Ms liu. Thanks.

Mr. Li: I'd like to start this interview with some questions. Why do you think you are qualified for this position?

Ms Liu: According to your advertisement, you want an experienced software engineer. I think my background meets the requirement of this position.

Mr. Li: Then tell me something about your background.

Ms Liu: My major was computer science when I was at college, and I am quite familiar with Visual C++ and Java language.

Mr. Li: Well, what do you think about the development in computers?

Ms Liu: The developments in software are going ahead very quickly and more and more problems are resolved by software. In some regions, the hardware is completely replaced by software. So I think the software industry has a great future.

Mr. Li: Have you ever designed any programs concerning network?

Ms Liu: Yes, I have designed some programs for the network with Visual C++ and I have passed the test for programmers — MCSE.

Mr. Li: Have you got anything to ask me?

Ms Liu: Yes, can you tell me what's my responsibility in this position?

Mr. Li: Yes, of course. You would be responsible for the development of software products.

Ms Liu: I see. This is my advantage.

Mr. Li: Good. Have you got any other questions?

Ms Liu: No.

Mr. Li: Ok, I will contact you in a week. See you.

Ms Liu: Thank you, bye-bye.

## New words & Expressions:

have an appointment with    与……有约会

be qualified for    胜任，有资格做

go ahead    前进，发展

be resolved by    被解决，通过……解决

software engineer    软件工程师

## ☞ Section 2  Reading Material:

# Instruction to the Computer Application Software

What is the computer application software, and how does it differ from other categories of software? This lesson introduces you to some examples of application software and how they are used.

The term "software" refers to the set of electronic program instructions or data a computer processor reads in order to perform a task or operation. In contrast, the term "hardware" refers to the physical components that you can see and touch, such as the computer hard drive, mouse, and keyboard.

Software can be categorized according to what it is designed to accomplish. There are two main types of software: systems software and application software.

Systems software includes the programs that are dedicated to managing the computer itself, such as the operating system, file management utilities, and disk operating system (or DOS). The operating system manages the computer hardware resources in addition to applications and data. Without systems software installed in our computers, we would have to type the instructions for everything we wanted the computer to do.

Application software, or simply applications, are often called productivity programs or end-user programs because they enable the user to complete tasks, such as creating documents, spreadsheets, databases and publications, doing online research, sending e-mail, designing graphics, running businesses, and even playing games. Application software is specific to the task it is designed for and can be as simple as a calculator application or as complex as a word processing application. When you begin creating a document, the word processing software has already set the margins, font style and size, and the line spacing for you. But you can change these settings, and you have many more formatting options available. For example, the word processor application makes it easy to add color, headings, and pictures or delete, copy, move, and change the document's appearance to suit your needs.

Microsoft Word is a popular word-processing application that is included in the software suite of applications called Microsoft Office. A software suite is a group of software applications with related functionality. For example, office software suites might include word processing, spreadsheet, database, presentation, and e-mail applications. Graphics suites such as Adobe Creative Suite include applications for creating and editing images, while Sony Audio Master Suite is used for audio production.

A Web browser, or simply browser, is an application specifically designed to locate, retrieve, and display content found on the Internet. By clicking a hyperlink or by typing the URL of a website, the user is able to view Web sites consisting of one or more Web pages. Browsers such as Internet Explorer, Mozilla Firefox, Google Chrome, and Safari are just a few of the many available to choose from.

## *New words & Expressions:*

application software 应用软件
system software 系统软件
information 信息
compiler 编译器
interpreter 解释器
reboot 重启
database 数据库
word processor 文字编辑器
spreadsheet 电子表格
run 运行

## *Exercises*

1. Answer the following questions according to the text.

(1) Which is also called end-user programs?

(2) What are Microsoft Office and its components?

(3) Please list the common browser.

2. Translate the following sentences into English/Chinese.

(1) Microsoft Word is a popular word-processing application that is included in the software suite of applications called Microsoft Office.

(2) The term "software" refers to the set of electronic program instructions or data a computer processor reads in order to perform a task or operation.

(3) 应用软件可以根据设计目的来进行分类。

3. Match the items in Column A with the translated versions in Column B.

| A | B |
|---|---|
| (1) system software | (　) a. 系统软件 |
| (2) application software | (　) b. 图形艺术软件 |
| (3) computer aided design | (　) c. 企业软件 |
| (4) enterprise software | (　) d. 计算机辅助设计 |
| (5) computer aided engineering | (　) e. 计算机辅助工程 |
| (6) graphic art software | (　) f. 应用软件 |

## ☞ Section 3　Reading Material：

# Using the Speak Feature in Office 2010

The Speak feature in Office 2010 enables text-to-speech in OneNote, Outlook, PowerPoint, and Word. By default, Speak is not present on the Ribbon, so you will need to add it to either the Ribbon or the Quick Access Toolbar. Note, you may want to use text-to-speech playback without turning on the Mini Translator or with the keyboard instead of the mouse.

Speak requires a text-to-speech engine matching the language of the text. You can see your installed text-to-speech engines in the Control Panel:

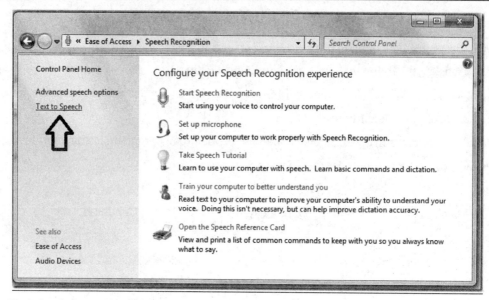

Let's begin by adding Speak to the Quick Access Toolbar:

1. Start by launching Word, and open the Backstage View by clicking on the File button.

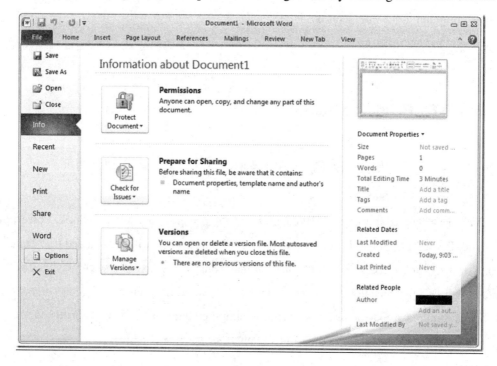

2. Click Options and navigate to the Quick Access Toolbar tab.
3. In the Choose commands from: drop-down menu, select Commands Not in the Ribbon.
4. Select Speak from the scroll box and click Add.

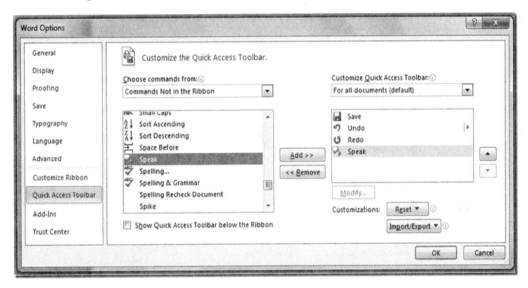

5. Click OK, and the Speak icon will now appear in the Quick Access Toolbar.

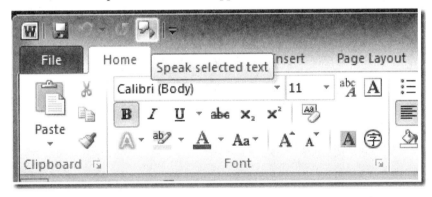

You can also add Speak to the Ribbon by using the following steps:

1. Start by launching Word, and open the Backstage View by clicking on the File button.
2. Select the Customize Ribbon tab in Options.
3. In the Choose commands from: drop-down menu, select Commands Not in the Ribbon.
4. Create a custom tab or a new group by clicking New Tab or New Group (You can rename

the Tab or the Group using the Rename button).

5. Select Speak from the scroll box and add it to your custom Tab and Group by clicking Add.

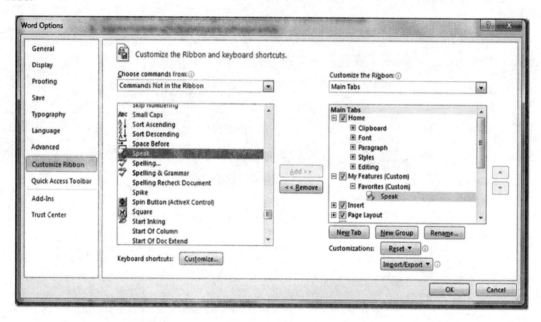

6. Click OK, and the Speak icon will now appear in the Ribbon.

Speak is now ready for text-to-speech playback, when the cursor is within a word or you have selected some text!

Click Speak to begin text-to-speech playback. Clicking the button during speech will cancel playback.

## *New words & Expressions:*

feature   特性
the control panel   控制面板
option   选项

## *Exercises*

1. Answer the following question according to the text.

How to use the Speak Feature in Office 2010?

2. Translate the following sentences into Chinese.

(1) By default, Speak is not present on the Ribbon, so you will need to add it to either the Ribbon or the Quick Access Toolbar.

(2) Speak requires a text-to-speech engine matching the language of the text.

3. Fill in the blank with the right words.

> infected, downloading, filter, scanner, executed, media

Computer viruses are programs that must be triggered or somehow 1._____ before they can infect your computer system and spread to others. Examples include opening a document 2._____ with a "macro virus", booting with a diskette infected with a "boot sector" virus, or double-clicking on an infected program file. Viruses can then be spread by sharing infected files on a diskette, network drive, or other 3._____, by exchanging infected files over the Internet via e-mail attachments, or by 4._____ questionable files from the Internet.

At the beginning of February, 2001, UH Information Technology Services installed an e-mail virus 5._____ on its mail gateway. This software is able to 6._____ out viruses sent as e-mail attachments before it is received by ITS customers. However, it is not foolproof and new and unrecognized viruses may still be able to get through the filter.

## ☞ Section 4  Extended Reading：

## The Office 2010 Suite will be Released

Microsoft Office has been the best selling piece of software for five years running. How does the company plan to compete when it releases Office 2010? Make most of it available for free.

To counter the popularity of Google's online Docs application, which bundles a word-processing tool, spreadsheet editing, presentation functionality and more into one free app, Microsoft is adding its own free online component: Web apps.

The Office 2010 suite, which Microsoft will release to business customers tomorrow and to consumers in early June, will include the ability to create, edit, view and share files online using the company's SkyDrive website. In fact, you won't even need to buy the program to use the online tools; Microsoft will make the majority of Office functionality available for free to anyone — whether they've bought the new suite or not.

To use the online functions, simply visit SkyDrive.com (or the Facebook-ified version at docs.com). You can also save a file from one of the Office 2010 apps directly to SkyDrive. Then visit the file from any browser, anywhere you go, and click the edit button to bring up editing options. The free versions of these apps don't include all of the functionality you'll find if you buy Office 2010, but most users will probably find it sufficient.

So which is easier to use, Google Docs or Office 2010? Both are relatively straightforward programs, though Office adds some neat extras, such as the ability to broadcast a PowerPoint presentation across the Internet. Office Web Apps have a few rough edges still, but once ironed out they'll be very robust programs — especially considering the price.

There's more to Office 2010 than just that, of course. Outlook fans will appreciate a new Social Connector feature, which brings the e-mail and calendaring program into the world of social networking. It lets you sync contact data with popular social networks, sharing status, pictures, shared documents and more.

Office 2010 also adds a very neat "broadcast" function for PowerPoint presentations. The feature uploads your presentation to a secure website and gives you a unique URL to it; pass the URL to your friends or colleagues to create an impromptu presentation from wherever you are. It's a very convenient way to collaborate — though to start such a presentation, you'll need to buy Microsoft Office 2010.

There are many other smaller improvements, of course, such as Excel's Sparklines — a new data visualization technique that adds little trend lines into individual cells — new text effects for Word, video editing functions in PowerPoint and so on.

But the biggest change users will encounter is hardly a new one at all. When Microsoft released Office 2007, the company rewrote the rules for interfacing software with "the Ribbon", a new user interface paradigm that bubbles up contextual commands — in theory, just the ones you'll want for whatever you're doing.

The interface is polarizing: You either love it or can't figure out how to use it. And that's part of the reason MS didn't have a gigantic hit on its hands with the last version of its productivity suite. With Office 2010, the Ribbon expands onto all of the Office apps. Learn to love it, in other words — and embrace the online future of Office.

Office 2010 sells in several versions. Office Professional, which includes Word 2010, Excel 2010, PowerPoint 2010, OneNote 2010, Outlook 2010, Publisher 2010, Access 2010, and premium technical support is priced at $499 for the full boxed copy or $349 for the product key card.

Office Home and Business sells for $279 in a box, or $199 for a product key that lets you download and activate the app online. This version includes Word 2010, Excel 2010, PowerPoint 2010, OneNote 2010, and Outlook 2010.

Office Home and Student is priced at $149 for the boxed version and $119 for the product key card and includes Word 2010, Excel 2010, PowerPoint 2010, OneNote 2010, and the Office Web Apps. It is available in a Family Pack, allowing use on three PCs in one home.

## ☞ Section 5  Extended Reading:

## Microsoft Teams Challenges Slack for Office Dominance

Chatting and collaborating with colleagues used to be all about watercoolers and whiteboards. These days it's more about videoconferencing, e-mail and at least one form of instant messaging.

Microsoft is the latest company to jump on the work-chat bandwagon with the launch of Teams at a New York event Wednesday. Teams brings together chats, notes and planning with the intention of being a standalone, catchall tool to promote good organization and productivity in offices, no matter what type of project people are working on.

The focus of the launch was on collaboration, bringing together the communications expertise of Microsoft's Skype division with the work-habits knowledge of its Office unit. Teams supports threaded chat, personalization and Office 365 integration. It also has companion apps for iOS, Android and Windows Phone.

"This is an experience that truly empowers that art form of how teams work and how teams drive success," said Microsoft CEO Satya Nadella.

Microsoft isn't the only company to show off a Slack rival in recent weeks. Last month at an event in London, Facebook unveiled Workplace by Facebook, a work-based collaboration and chat room platform. The success of Slack, which boasts more than 3 million daily users at last count, shows there's huge demand for such services.

Unlike with Slack, in Teams you can reply to messages by creating threads.

Microsoft hopes to set Teams apart with features that bring Office 365 tools into the service. You can autosave attachments to folders for each stream, pin frequently used files to your dashboard, integrate your planner or task manager and use third-party plug-ins like Zendesk.

Robots also play a key role in the operation of Teams, including T-bot, an intelligent assistant, and Who Bot, which is designed to answer questions about who does what in your organization.

Microsoft is rumoured to have mulled buying Slack earlier this year for $8 billion (the company declined to comment on this at the time), but neither Nadella nor company founder and advisor Bill Gates was reportedly convinced it was a smart buy, preferring instead to focus efforts on its own Skype.

On Wednesday, in a full-page ad it took out in *The New York Times* and also published online, Slack acknowledged it now has "some competition", and it spoke to its rivals.

"It's validating to see you've come around to the same way of thinking," the company said, "We want to give you some friendly advice."

The ad was Slack's attempt to remind the world why it has become so successful. But the fact it was put out there in the first place may also suggest the company is feeling jittery at the challenge from Microsoft.

A preview of Teams is available immediately in 181 countries in 18 languages. Microsoft plans to release the full version sometime during the first three months of next year.

## ☞ Section 6  Extended Reading:

## The Way Ahead: Innovating Together in China
## —by William H. Gates

When I visited here in 1997—10 years ago—I was very impressed by the talent, the enthusiasm and the creativity of the students that I met at Tsinghua. And it inspired me to support Microsoft in creating a research lab here in Beijing.

That research lab has gone on to incredible success, led by Harry Shum and joined by top university graduates from this school and others. It's made huge contributions to Microsoft. And if you look at various conferences getting together to discuss the state-of-the-art issues, the researchers from this lab are making huge contributions. Or if you look at the products, even the recent Windows Vista or Office 2007, we have substantial innovations in those products.

This is an incredible time to be a student at this university. The frontiers of science, including computer science, have never been nearer. The opportunity to improve people's lives in every way has never been stronger.

We've really just scratched the surface of the digital revolution. Yes, we have about a billion personal computers that are connected up to the Internet. And we've already started to transform the way people think studying information and sharing information. But there are so much more that we can do.

The exponential improvement in not only the processor transistors but also what we have in storage or optic fiber bandwidth give us an opportunity to apply software that can work in a far more powerful way.

For example, when we think about TV today, it's just a passive, non-interactive situation, but that's in the process of changing: changing so that you can get any show that you want on the Internet, changing so that it can be interactive so that you can learn as you go home, changing so that the flexibility even to talk and collaborate with others will be part of that experience.

If you think about product design—and products are going to be designed digitally. I spent several hours yesterday at the Agricultural Rice Institute talking with the experts there about how they are using software technology to sequence different rice varieties. And they're optimistic that they can come up with new varieties that will require less fertilizer, less water and yet increase the yield. And its advances like that really reach out and change the lives not just of those of us who work in technology but people everywhere.

The advances in medicine are dependent on software, software that can manage the databases and understand the complex systems. I'm very optimistic that we can make big breakthroughs based on what software will provide.

And if you think about your mobile phone going from being just a voice device to being something that can be a "digital wallet", that can show you maps and you'll be able to talk to it and ask for information and it will go out to the Internet and find the things that you're interested in.

In terms of learning, by creating what we call the "student tablet" that will be very inexpensive and the size of a tablet but wirelessly connected to the Internet and able to record your voice or recognize your ink handwriting and yet provide learning experiences that are far more effective — and in fact bringing together all of the world's knowledge on the Internet in a very attractive form. Teachers will be able to see the world's best lectures, and they'll be able to see the best materials and for the first time start to share with each other. And so for anyone who wants to learn or wants to teach, it will be a very, very different world.

We certainly have some very tough and interesting problems that I know the students here will be making breakthroughs in. Writing software that's reliable, that's totally secure, software that can handle parallel execution, software that's very easy to use and software that can solve some very tough problems, for example, problems of artificial intelligence that we've already spent many decades working on. And so this is an amazing time to be working in the sciences and particularly in computer science.

It's also an amazing time to this country. What's going on in China and the growth of its economy with incredible contributions not just within the country but to the global economy as well — you know, starting to be a very major contributor in all the sciences, advances in the medical drugs, advances in computer science. China will start to play a very substantial role, and part of that is the investments that have been made in having world-class universities, of which Tsinghua is really the shining example.

For Microsoft, we have a commitment to work with our partners here and make them successful, to make sure that there are literally hundreds of software start-ups that not only sell in the market here but sell to the entire world.

We also want to make sure that the digital advances are available to all the citizens. And so whether it's displaced workers or migrants or people who have disabilities — for example, blind people — trying to use the computer, there are these special programs that we can put together to make sure that software really isn't just for the few but really is about empowering everyone. A good example of that is the 170 Hope Cyber Schools where we provided lots of training.

This commitment to think long-term and this commitment to the future is something that the Chinese Government, this University and Microsoft all share. And so the opportunity of working together on new curriculum, faculty exchange: these have strengthened all three of our institutions.

In the last several years, over 2,000 students from 100 Asian universities have worked at our research center here and we've awarded over 170 fellowships, so it's really become a mixing ground for the most talented people in the region. And of course, the university that has the largest representation has been Tsinghua.

Many more researchers are teaching courses, including the course here called "Hot Topics in Computing Research" that I think is a very novel type of course and I think will be a model for many other universities as well.

Today I get a chance to announce some new program between Microsoft and Tsinghua, which is the "Microsoft Distinguished Visiting Professorship" at Tsinghua. And under this program, our research group will support a world-renowned computer scientist every year to visit the University at Professor Yao's Institute for Theoretical Computer Science. And the first recipient of this — a very impressive recipient — is Professor Frans Kaashoek from MIT.

And so I hope you get a sense of my optimism — optimism about what software can do and the interesting breakthroughs we can all make, opportunities to use that — use it for making businesses more effective, making jobs more interesting, designing far better jobs than ever before but also using it for the handicapped, for education, for outreach so that any student who has an opportunity to connect up to the Internet will have the same type of opportunities as most privileged students.

So these are amazing times and, you know, I think the intersection of what's going on in China, what's going on with companies like Microsoft to take this longterm approach and the great academic tradition that is exemplified by the excellence of this university, I think, we can all have a very high expectation. And certainly we, recommitted to working with all of you to realize that potential. Thank you!

# Unit 5

## Programming Language

After reading this unit and completing the exercises, you will be able to
- Understand what a programming language is.
- Be familiar with the history of programming language.
- Identify all the factors that should be considered when selecting a programming language.
- Apply your knowledge when using programming language.

☞ **Section 1  Situational Dialogue:**

### Call about Courses of Computer Programming

A: Hello, this is the admissions office. Can I help you?

B: Hi. I'm calling about your continuing education program.

A: What would you like to know?

B: I want to become certified in computer programming. Do you offer any part-time courses for adult further education?

A: Yes, we offer both night and weekend courses in a number of different subjects.

B: How do I sign up?

A: If you give me your address, I can mail you an information packet and the application forms.

B: Great!

## New words & Expressions:

call about    打电话谈……
computer programming    计算机编程
sign up    报名参加
admissions office    招生办公室

## ☞ Section 2   Reading Material：

# Introduction to Programming Language

A programming language is used to write computer programs including applications, utilities, and systems programs. Before the Java and C# programming languages appeared, computer programs were either compiled or interpreted.

A compiled program is written as a series of humanly understandable computer instructions that can be read by a compiler and linker and translated into machine code so that a computer can understand and run it. Fortran, Pascal, Assembly Language, C, and C++ programming languages are almost always compiled in this way. Other programs, such as Basic, JavaScript, and VBScript, are interpreted. The differences between compiled and interpreted languages can be confusing.

**Compiling a Program**

The development of a compiled program follows these basic steps:

1. Write or edit the program.
2. Compile the program into machine code files that are specific to the target machine.
3. Link the machine code files into a runnable program (known as an EXE file).
4. Debug or run the program.

Interpreting a Program:

Interpreting a program is a much faster process that's helpful for novice programmers when editing and testing their code. These programs run slower than compiled programs.

The steps to interpret a program are:

1. Write or edit the program.

2. Debug or run the program using an interpreter program.

**Java and C#**

Both Java and C# are semi-compiled. Compiling Java generates bytecode that is later interpreted by a Java virtual machine. As a result, the code is compiled in a two-stage process.

C# is compiled into Common Intermediate Language, which is then run by the Common Language Runtime part of the NET framework, an environment that supports just-in-time compilation.

The speed of C# and Java is almost as fast as true compiled language. As far as speed goes, C, C++, and C# all are sufficiently speedy for games and operating systems.

Are There Many Programs on a Computer?

From the moment you turn on your computer, it is running programs, carrying out instructions, testing RAM and accessing the operating system on its drive.

Each and every operation that your computer performs has instructions that someone had to write in a programming language. For example, the Windows 10 operating system has roughly 50 million lines of code. These had to be created, compiled and tested—a long and complex task.

What Programming Languages Are Now In Use?

Top programming languages for PCs are Java and C++ with C# close behind and C holding its own. Apple products use Objective-C and Swift programming languages.

There are hundreds of small programming languages out there, but other popular programming languages include:

- Python
- PHP
- Perl
- Ruby
- Go
- Rust

- Scala

There have been many attempts to automate the process of writing and testing programming languages by having computers write computer programs, but the complexity is such that, for now, humans still write and test computer programs.

**The future for programming languages**

Computer programmers tend to use programming languages they know. As a result, the old tried-and-true languages have hung around for a long time. With the popularity of mobile devices, developers may be more open to learning new programming languages. Apple developed Swift to eventually replace Objective-C, and Google developed Go to be more efficient than C. Adoption of these new programs has been slow, but steady.

## *New words & Expressions:*

programming language 编程语言
keyword 关键字
communicate with 与……通信
command 指令
procedural language 过程语言
object-oriented programming 面向对象编程
translate ... into ... 把……翻译……
assembly language 汇编语言
programmer 程序员
compiler 编译器
loader 装载器
linker 链接器

## *Exercises*

1. Answer the following questions according to the text.

(1) Is JavaScript the future of programming?

(2) Talk about the categories of programming languages.

(3) Please list the common programming languages.

2. Translate the following sentences into English/Chinese.

(1) A compiled program is written as a series of humanly understandable computer instructions that can be read by a compiler and linker and translated into machine code so that a computer can understand and run it.

(2) Each and every operation that your computer performs has instructions that someone had to write in a programming language.

(3) 编译性语言和解释性语言的区别让人迷惑。

3. Match the items in Column A with the translated versions in Column B.

| A | B |
|---|---|
| (1) grammatical rules | (  ) a. 执行 |
| (2) operand | (  ) b. 机器语言 |
| (3) carry out | (  ) c. 操作数 |
| (4) programming language | (  ) d. 编程语言 |
| (5) primary memory | (  ) e. 语法规则 |
| (6) order of magnitude | (  ) f. 二进制数值 |
| (7) machine language | (  ) g. 主存 |
| (8) assembly language | (  ) h. 汇编语言 |
| (9) implementation | (  ) i. 数量级 |
| (10) binary number | (  ) j. 实施 |

## ☞ Section 3  Reading Material:

# The C Programming Language

1. Introduction to the basics of C Programming

The C programming language is a popular and widely used programming language for creating computer programs. Programmers around the world embrace C because it gives maximum control and efficiency to the programmer.

If you are a programmer, or if you are interested in becoming a programmer, there are a couple of benefits you gain from learning C:

You will be able to read and write code for a large number of platforms — everything from microcontrollers to the most advanced scientific systems can be written in C, and many modern operating systems are written in C.

The jump to the object oriented C++ language becomes much easier. C++ is an extension of C, and it is nearly impossible to learn C++ without learning C first.

2. What is C?

C is a computer programming language. That means that you can use C to create lists of instructions for a computer to follow. C is one of thousands of programming languages currently in use. C has been around for several decades and has won widespread acceptance because it gives programmers maximum control and efficiency. C is an easy language to learn. It is a bit more cryptic in its style than some other languages, but you get beyond that fairly quickly.

C is what is called a compiled language. This means that once you write your C program, you must run it through a C compiler to turn your program into an executable that the computer can run (execute). The C program is the human-readable form, while the executable that comes out of the compiler is the machine-readable and executable form. What this means is that to write and run a C program, you must have access to a C compiler. If you are using a UNIX machine (for example, if you are writing CGI scripts in C on your host's UNIX computer, or if you are a student working on a lab's UNIX machine), the C compiler is available for free. It is called either "cc" or "gcc" and is available on the command line.

If you are a student, then the school will likely provide you with a compiler — find out what the school is using and learn about it. If you are working at home on a Windows machine, you are going to need to download a free C compiler or purchase a commercial compiler. A widely used commercial compiler is Microsoft's Visual C++ environment (it compiles both C and C++ programs). Unfortunately, this program costs several hundred dollars. If you do not have hundreds of dollars to spend on a commercial compiler, then you can use one of the free compilers available on the Web.

3. The Simplest C Program

Let's start with the simplest possible C program and use it both to understand the basics of C and the C compilation process. Type the following program into a standard text editor (vi or emacs on UNIX, Notepad on Windows or TeachText on a Macintosh). Then save the program to a file named samp.c. If you leave off .c, you will probably get some sorts of error when you compile it, so make sure you remember the .c. Also, make sure that your editor does not automatically append some extra characters (such as .txt) to the name of the file. Here's the first program:

```
#include <stdio.h>
int main()
{
    printf("Hello world!\n");
    return 0;
}
```

When executed, this program instructs the computer to print out the line "Hello world!"—then the program quits. You can't get much simpler than that!

To compile this code, take the following steps:

On a UNIX machine, type gcc samp.c -o samp (if gcc does not work, try cc). This line invokes the C compiler called gcc, asks it to compile samp.c and asks it to place the executable file it creates under the name samp. To run the program, type samp (or, on some UNIX machines, ./samp).

On a DOS or Windows machine using DJGPP, at an MS-DOS prompt type gcc samp.c -o samp.exe. This line invokes the C compiler called gcc, asks it to compile samp.c and asks it to place the executable file it creates under the name samp.exe. To run the program, type samp.

## *New words & Expressions:*

platform    平台
extension   扩展

executable  可执行的
download  下载

## Exercises

1. Answer the following questions according to the text.
(1) Both C and C++ are object oriented languages.
(2) Must every C program have a function named main at the beginning in the cod?
(3) Please write a simple c program.
2. Translate the following sentences into English/Chinese.
(1) C语言是一门用于编写计算机程序的流行而且广泛使用的语言。

(2) C is one of thousands of programming languages currently in use. C has been around for several decades and has won widespread acceptance because it gives programmers maximum control and efficiency.

(3) If you are working at home on a Windows machine, you are going to need to download a free C compiler or purchase a commercial compiler.

3. Fill in the blank with the right words.

> efficiency, programming, programmers, compiler, executable,
> machine-readable, instructions, compiled, human-readable, control

C is a computer _____ language. That means that you can use C to create lists of _____ for a computer to follow. C is one of thousands of programming languages currently in use. C has been around for several decades and has won widespread acceptance because it gives _____ maximum _____ and _____.

C is what is called a _____ language. This means that once you write your C program, you must run it through a C _____ to turn your program into an _____ that the computer can run. The C program is the _____ form, while the executable that comes out of the compiler is the _____ and executable form. What this means is that to write and run a C program, you must have access to a C compiler.

## Section 4  Extended Reading:

## What is the "Java language"?

Java is a general-purpose computer-programming language that is concurrent, class-based, object-oriented, and specifically designed to have as few implementation dependencies as possible. It is intended to let application developers "write once, run anywhere" (WORA), meaning that compiled Java code can run on all platforms that support Java without the need for recompilation. Java applications are typically compiled to bytecode that can run on any Java virtual machine (JVM) regardless of computer architecture. As of 2016, Java is one of the most popular programming languages in use, particularly for client-server web applications, with a reported 9 million developers.  Java was originally developed by James Gosling at Sun Microsystems (which has since been acquired by Oracle Corporation) and released in 1995 as a core component of Sun Microsystems' Java platform. The language derives much of its syntax from C and C++, but it has fewer low-level facilities than either of them.

The original and reference implementation Java compilers, virtual machines, and class libraries were originally released by Sun under proprietary licenses. As of May 2007, in compliance with the specifications of the Java Community Process, Sun relicensed most of its Java technologies under the GNU General Public License. Others have also developed alternative implementations of these Sun technologies, such as the GNU Compiler for Java (bytecode compiler), GNU Classpath (standard libraries), and IcedTea-Web (browser plugin for applets).

The latest version is Java 10, released on March 20, 2018,  which follows Java  9 after only six months  in line with the new release schedule. Java 8 is still supported but there will be no more security updates for Java 9. Versions earlier than Java 8 are supported by companies on a commercial basis; e.g. by Oracle back to Java 6 as of October 2017 (while they still "highly recommend that you uninstall" pre-Java 8 from at least Windows computers).

## ☞ Section 5  Extended Reading:

# Object-oriented Programming

Object Oriented Programming (OOP) is a term loosely applied to mean any kind of programming that uses a programming language with some object oriented constructs or programming in an environment where some object oriented principles are followed. It is a programming paradigm that uses "objects" – data structures consisting of datafields and methods together with their interactions – to design applications and computer programs.

Object-oriented programming has roots that can be traced to the 1960s. As hardware and software became increasingly complex, manageability often became a concern. Researchers studied ways to maintain software quality and developed object-oriented programming in part to address common problems by strongly emphasizing discrete, reusable units of programming logic.

The technology focuses on data rather than processes, with programs composed of self-sufficient modules ("classes"), each instance of which ("objects") contains all the information needed to manipulate its own data structure ("members"). This is in contrast to the existing modular programming which had been dominant for many years that focused on the function of a module, rather than specifically the data, but equally provided for code reuse, enabling collaboration through the use of linked modules (subroutines). This more conventional approach, which still persists, tends to consider data and behavior separately. Object-oriented programming was not commonly used in mainstream software application development until the early 1990s. Many modern programming languages now support OOP.

One of the principal advantages of object-oriented programming techniques over procedural programming techniques is that they enable programmers to create modules that do not need to be changed when a new type of object is added. A programmer can simply create a new object that inherits many of its features from existing objects. Other pieces of software can access the object only by calling its functions and procedures that have been allowed to be called by outsiders. This makes object-oriented programs easier to modify.

An object-oriented program may thus be viewed as a collection of interacting objects, as opposed to the conventional model, in which a program is seen as a list of tasks (subroutines) to perform. In OOP, each object is capable of receiving messages, processing data, and sending messages to other objects and can be viewed as an independent "machine" with a distinct role or responsibility.

There are some important principles of object-oriented programming below:

Class, Object, Message, Inheritance.

**Class**

A class is a blueprint that defines the variables and the methods common to all objects of a certain kind and is used to manufacture or create objects. It defines the abstract characteristics of a thing (object). For example, the class Car would consist of traits shared by all cars, such as brand and color (characteristics), and the ability to move and moo (behaviors). Classes provide modularity and structure in an object-oriented computer program. A class should typically be recognizable to a non-programmer familiar with the problem domain, meaning that the characteristics of the class should make sense in context. Collectively, the properties and methods defined by a class are called "members".

**Object**

An object is also known as an instance. It is a software bundle of related state and behavior. It is an instance (that is, an actual example) of a class. Software objects are often used to model the real-world objects that you find in everyday life. It consists of state and the behavior that's defined in the object's class. For example, the class car provides a pattern or blueprint for car objects by listing the characteristics and behaviors they can have; the object car with plate number "皖 C35828" is one particular car.

**Message**

Software objects interact and communicate with each other by sending messages to each other. When object A wants object B to perform one of B's methods, object A sends a message to object B. For example, the object called driver may fire a car to start up by passing a "fire" message which invokes the car's "fire" method.

**Inheritance**

Inheritance is a way to form new classes using classes that have already been defined. Inheritance is employed to help reuse existing code with little or no modification. The new classes, known as sub-classes, inherit attributes and behavior of the pre-existing classes, which are referred to as super-classes. The inheritance relationship of sub-classes and super-classes gives rise to a hierarchy. For example, the class Car might have sub-classes called Lorry, Motorcycle, and Carriage. Suppose the Car class defines a method called brake and a property called wheelbase. Each of its sub-classes (Lorry, Motorcycle, and Carriage) will inherit these members, meaning that the programmer only needs to write the code for them once.

## ☞ Section 6  Extended Reading:

# How to Learn a Programming Language

If you have an interest in creating computer programs, mobile apps, websites, games or any other piece of software, you'll need to learn how to program. Programs are created through the use of a programming language. This language allows the program to function with the machine it is running on, be it a computer, a mobile phone, or any other piece of hardware.

**Choosing a language**

Determine your area of interest. You can start learning with any programming language (though some are definitely "easier" than others), so you'll want to start by asking yourself what it is you want to accomplish by learning a programming language. This will help you determine what type of programming you should pursue, and provide you a good starting point.

If you want to get into web development, you'll have a whole different set of languages that you'll need to learn as opposed to developing computer programs. Mobile app developing requires a different skillset than machine programming. All of these decisions will influence your direction.

Consider starting with a "simpler" language. Regardless of your decision, you may want to consider starting with one of the high-level, simpler languages. These languages are especially useful for beginners, as they teach basic concepts and thought processes that can apply to virtually any language.

The two most popular languages in this category are Python and Ruby. These are both object-oriented web application languages that use a very readable syntax.

"Object-oriented" means that the language is built around the concepts of "objects", or collections of data, and their manipulation. This is a concept that is used in many advanced programming languages such as C++, Java, Objective-C, and PHP.

Read through some basic tutorials for a variety of languages. If you're still not sure which language you should start learning, read through some tutorials for a few different languages. If one language makes a bit more sense than the others, try it out for a bit to see if it clicks. There are countless tutorials for every programming available online, including many on wikiHow:

• Python — A great starter language that is also quite powerful when you get familiar with it. Used for many web applications and a number of games.

• Java — Used in countless types of programs, from games to web applications to ATM software.

• HTML — An essential starting place for any web developer. Having a handle on HTML is vital before moving on to any other sort of web development.

• C — One of the older languages, C is still a powerful tool, and is the basis for the more modern C++, C#, and Objective-C.

**Starting small**

Learn the core concepts of the language.

While the parts of this step that apply will vary depending on the language you choose, all programming languages have fundamental concepts that are essential to building useful programs. Learning and mastering these concepts early will make it easier to solve problems and create powerful and efficient code. Below are just some of the core concepts found in many different languages:

Variables — A variable is a way to store and refer to changing pieces of data. Variables can be manipulated, and often have defined types such as "integers", "characters", and others, which determine the type of data that can be stored. When coding, variables typically have names that make them somewhat identifiable to a human reader. This makes it easier to understand how the variable interacts with the rest of the code.

Conditional Statements — A conditional statement is an action that is performed based on whether the statement is true or not. The most common form of a conditional statement is the "If-Then" statement. If the statement is true (e.g. x = 5) then one thing happens. If the statement is false (e.g. x != 5), then something else happens.

Functions or Subroutines — The actual name for this concept may be called something different depending on the language. It could also be "Procedure", a "Method", or a "Callable Unit". This is essentially a smaller program within a larger program. A function can be "called" by the program multiple times, allowing the programmer to efficiently create complex programs.

Data input — This is a broad concept that is used in nearly every language. It involves handling a user's input as well as storing that data. How that data is gathered depend on the type of program and the inputs available to the user (keyboard, file, etc.). This is closely linked to Output, which is how the result is returned to the user, be it displayed on the screen or delivered in a file.

Install any necessary software. Many programming languages require compilers, which are programs designed to translate the code into a language that the machine can understand. Other languages, such as Python, use an interpreter which can execute the programs instantly without compiling.

Some languages have IDEs (Integrated Development Environment) which usually contain a code editor, a compiler and/or interpreter, and a debugger. This allows the programmer to perform any necessary function in one place. IDEs may also contain visual representations of object hierarchies and directories.

There are a variety of code editors available online. These programs offer different ways of highlighting syntax and provide other developer-friendly tools.

**Creating your first program**

Focus on one concept at a time. One of the first programs taught for any language is the "Hello World" program. This is a very simple program that displays the text "Hello, World" (or some variation), on the screen. This program teaches first-time programmers the syntax to create a basic, functioning program, as well as how to handle displaying output. By changing the text, you can learn how basic data is handled by the program. Below are some wikiHow guides on creating a "Hello World" program in various languages:

- Hello World in Python
- Hello World in Ruby
- Hello World in C
- Hello World in PHP
- Hello World in C#
- Hello World in Java

Learn through deconstruction of online examples. There are thousands of code examples online for virtually every programming languages. Use these examples to examine how various aspects of the language work and how different parts interact. Take bits and pieces from various examples to create your own programs.

Examine the syntax. The syntax is the way the language is written so that the compiler or interpreter can understand it. Each language has a unique syntax, though some elements may be shared across multiple languages. Learning the syntax is essential for learning how to program in the language, and is often what people think of when they think about computer programming. In reality, it is simply the foundation upon which more advanced concepts are built.

Experiment with changes. Make changes to your example programs and then test the result. By experimenting, you can learn what works and what doesn't much quicker than by reading a book or guide. Don't be afraid to break your program; learning to fix errors is a major part of any development process, and new things almost never work right the first time.

Start practicing debugging. When you're programming, you're invariably going to come across bugs. These are errors in the program, and can manifest virtually anywhere. Bugs can be harmless quirks in the program, or they can be major errors that keep the program from compiling or running. Hunting down and fixing these errors is a major process in the software development cycle, so get used to doing this early.

As you experiment with changing basic programs, you're going to come across things that don't work. Figuring out how to take a different approach is one of the most valuable skills you can have as a programmer.

Comment all of your code. Nearly all programming languages have a "comment" function that allows you to include text that is not processed by the interpreter or compiler. This allows you to leave short, but clear, human-language explanations of what the code does. This will not only help you remember what your code does in a large program, it is an essential practice in a collaborative environment, as it allows others to understand what your code is doing.

**Practicing regularly**

Code daily. Mastering a programming language takes time above all else. Even a simpler language like Python, which may only take a day or two to understand the basic syntax, takes lots of time to become truly proficient at. Like any other skill, practice is the key to becoming more proficient. Try to spend at least some time each day coding, even if it's only for an hour between work and dinner.

Set goals for your programs. By setting attainable but challenging goals, you will be able to start solving problems and coming up with solutions. Try to think of a basic application, such as a calculator, and develop a way to make it. Use the syntax and concepts you've been learning and apply them to practical uses.

Talk with others and read other programs. There are lots of programming communities dedicated to specific languages or disciplines. Finding and participating in a community can do wonders for your learning. You will gain access to a variety of samples and tools that can aid you in your learning process. Reading other programmers' code can inspire you and help you grasp concepts that you haven't mastered yet.

Check out programming forums and online communities for your language of choice. Make sure to participate and not just constantly ask questions. These communities are usually viewed as a place of collaboration and discussion and not simply Q&A. Feel free to ask for help, but be prepared to show your work and be open to trying different approaches.

Once you have some experience under your belt, consider attending a hack-a-thon or programming jam. These are events where individuals or teams compete against the clock to develop a functional program, usually based around a specific theme. These events can be a lot of fun and are a great way to meet other programmers.

Challenge yourself to keep it fun. Try to do things that you don't know how to do yet. Research ways to accomplish the task (or a similar one), and then try to implement that in your own program. Try to avoid being content with a program that "basically" works; do everything you can to make sure every aspect works flawlessly.

**Expanding your knowledge**

Take a few training courses. Many universities, community colleges, and community centers offer programming classes and workshops that you can attend without having to enroll in the school. These can be great for new programmers, as you can get hands-on help from an experienced programmer, as well as network with other local programmers.

Buy or borrow a book. There are thousands of instructional books available for every conceivable programming language. While your knowledge should not come strictly from a book, they make great references and often contain a lot of good examples.

Study math and logic. Most programming involves basic arithmetic, but you may want to study more advanced concepts. This is especially important if you are developing complex simulations or other algorithm-heavy programs. For most day-to-day programming, you don't need much advanced math. Studying logic, especially computer logic, can help you understand how best to approach complex problem solving for more advanced programs.

Never stop programming. There is a popular theory that becoming an expert takes at least 10,000 hours of practice. While this is up for debate, the general principle remains true: mastery takes time and dedication. Don't expect to know everything overnight, but if you stay focused and continue to learn, you may very well end up an expert in your field.

Learn another programming language. While you can certainly get by with mastering one language, many programmers help their chances of success in the field by learning multiple languages. Their second or third languages are usually complementary to their first one, allowing them to develop more complex and interesting programs. Once you have a good grasp on your first program, it may be time to start learning a new one.

You will likely find that learning your second language goes much quicker than the first. Many core concepts of programming carry over across languages, especially if the languages are closely related.

**Applying your skills**

Enroll in a four-year program. While not strictly necessary, a four-year program at a college or university can expose you to a variety of different languages, as well as help you network with professionals and other students. This method certainly isn't for everyone, and plenty of successful programmers never attended a four-year institution.

Create a portfolio. As you create programs and expand your knowledge, make sure that all of your best work is saved in a portfolio. You can show this portfolio to recruiters and interviewers as an example of the work you do. Make sure to include any work done on your own time, and ensure that you are allowed to include any work done with another company.

Do some freelance work. There is a very large freelance market for programmers, especially mobile app developers. Take on a few small freelance jobs to get a feel for how commercial programming works. Oftentimes you can use the freelance jobs to help build your portfolio and point to published work.

Develop your own freeware or commercial programs. You don't have to work for a company to make money programming. If you have the skills, you can develop software yourself and release it for purchase, either through your own website or through another marketplace. Be prepared to be able to provide support for any software you release for commercial sale, as customers will expect their purchase to work.

Freeware is a popular way to distribute small programs and utilities. The developer doesn't receive any money, but it's a great way to build name recognition and make yourself visible in the community.

# Unit 6

## Network and Internet

After reading this unit and completing the exercises, you will be able to

- ◆ Understand the different kinds of network.
- ◆ Be familiar with the common network components.
- ◆ Describe what is the difference between Network and Internet.
- ◆ List the common applications of Internet.

☞ **Section 1  Situational Dialogue:**

### About Going Online

A: Can I use your laptop for a while?

B: Sure, go ahead.

A: Oh, isn't your computer Wi-Fi capable?

B: Yes, it is. You want go online? There are no wi-fi hotspots around.

A: Oh my, no internet access is killing me.

B: Can't you wait till you get home? Then you can surf the internet using the broadband, wireless connection or whatever you like.

A: No, I'm not feeling myself. I just want to check my e-mails, visit my favorite websites and chat with my friends.

B: Now I see, you must be suffering from discomgoogolation.

A: What does that mean? There's nothing wrong with me.

B: Well, the term "discomgoogolation" comes from "discombobulate" and "google". Because floods of information are just a mouse click away, net users are very likely to become addicted to the web.

A: That's alright. I just can't bear losing track of all the latest information. It almost drives me crazy.

B: Then, you're probably addicted.

## *New words & Expressions:*

wi-fi hotspot    无线热点
broadband    宽带
discomgoogolation    谷歌依赖症

### ☞ Section 2  Reading Material：

## Instruction to the Network

A network is a group of two or more computer systems linked together. There are many types of computer networks, including local-area networks(LANs): The computers are geographically close together (that is, in the same building).

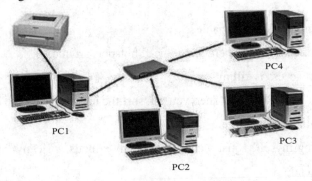

LAN

# Unit 6  Network and Internet

A local-area network(LAN) is a computer network that spans a relatively small area. Most LANs are confined to a single building or group of buildings.

LANs are capable of transmitting data at very fast rates, much faster than data can be transmitted over a telephone line; but the distances are limited, and there is also a limit on the number of computers that can be attached to a single LAN.

Wide-area networks (WAN): The computers are farther apart and are connected by telephone lines or radio waves. A computer network that spans a relatively large geographical area. Typically, a WAN consists of two or more local-area networks(LANs). Computers connected to a wide-area network are often connected through public networks, such as the telephone system. They can also be connected through leased lines or satellites. The largest WAN in existence is the Internet.

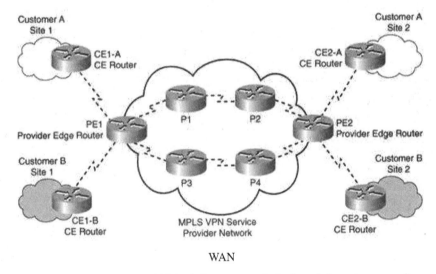

WAN

Metropolitan-area networks(MANs): A data network designed for a town or city.

In terms of geographic breadth, MANs are larger than local-area networks(LANs), but smaller than wide-area networks(WANs). MANs are usually characterized by very high-speed connections using fiber optical cable or other digital media.

**The other categories of networks**

In addition to these types, the following characteristics are also used to categorize different types of networks:

topology: The geometric arrangement of a computer system. Common topologies include a bus, a star, a ring and so on.

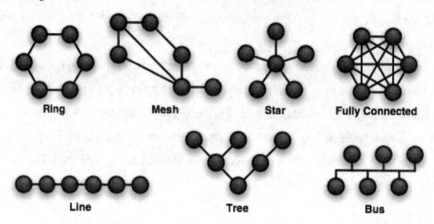

Types of network topologies

protocol: The protocol defines a common set of rules and signals that computers on the network use to communicate. One of the most popular protocols for LANs is called Ethernet. Another popular LAN protocol for PCs is the IBM token-ring network.

architecture: Networks can be broadly classified as using either a peer-to-peer or client/server architecture. Computers on a network are sometimes called nod.

## *New words & Expressions:*

computer network　计算机网络

leased line　租用线路

geographical　地理的

fiber optical cable　光纤

network topology diagram　网络拓扑图

to ken-ring network　令牌环网

topology　拓扑结构

architecture　体系结构

peer-to-peer  对等网

client / server architecture  客户端/服务器体系结构

Internet protocol suite  互联网协议族

hypertext  超文本

routing table  路由表

default route  默认路由

Border Gateway Protocol  边界网关协议

## *Exercises*

1. Answer the following questions according to the text.

(1) What are the types of networks?

(2) What is the difference between LAN and WAN?

(3) Please generally discuss about how the computer evolved from mechanical devices to electronic digital devices.

2. Translate the following sentences into English.

(1) 局域网(LAN)是指计算机网络跨越一个相对较小的区域。

(2) 常见的拓扑结构包括总线型、星型和环型。

## ☞ Section 3  Reading Material：

## Instruction to the Internet

The Internet is a global system of interconnected computer networks that use the standard Internet protocol suite(TCP/IP) to link several billion devices worldwide. It is a network of networks that consists of millions of private, public, academic, business, and government networks, of local to global scope, that are linked by a broad array of electronic, wireless, and optical networking technologies. The Internet carries an extensive range of information resources and services, such as the inter-linked hypertext documents and applications of the World Wide Web(WWW), the infrastructure to support email, and peer-to-peer networks for file sharing and telephony.

**Protocol**

In information technology, a protocol is the special set of rules that end points in a telecommunication connection use when they communicate. Protocols specify interactions between the communicating entities.

Protocols exist at several levels in a telecommunication connection. For example, there are protocols for the data interchange at the hardware device level and protocols for data interchange at the application program level. In the standard model known as Open Systems Interconnection(OSI), there are one or more protocols at each layer in the telecommunication exchange that both ends of the exchange must recognize and observe. Protocols are often described in an industry or international standard.

The TCP/IP Internet protocols, a common example, consist of:

• Transmission Control Protocol(TCP), which uses a set of rules to exchange messages with other Internet points at the information packet level.

• Internet Protocol(IP), which uses a set of rules to send and receive messages at the Internet address level.

• Additional protocols include the Hypertext Transfer Protocol(HTTP) and File Transfer Protocol(FTP), each with defined sets of rules to use with corresponding programs elsewhere on the Internet.

There are many other Internet protocols, such as the Border Gateway Protocol (BGP) and the Dynamic Host Configuration Protocol(DHCP).

**Routing**

Internet service providers connect customers, which represent the bottom of the routing hierarchy, to customers of other ISPs via other higher or same-tier networks.

At the top of the routing hierarchy are the Tier 1 networks, large telecommunication companies that exchange traffic directly with each other via peering agreements. Tier 2 and lower level networks buy Internet transit from other providers to reach at least some parties on the global Internet, though they may also engage in peering. An ISP may use a single upstream provider for connectivity, or implement multi homing to achieve redundancy and load balancing. Internet exchange points are major traffic exchanges with physical connections to multiple ISPs.

Computers and routers use a routing table in their operating system to direct IP packets to the next-hop router or destination. Routing tables are maintained by manual configuration or automatically by routing protocols. End-nodes typically use a default route that points toward an ISP providing transit, while ISP routers use the Border Gateway Protocol to establish the most efficient routing across the complex connections of the global Internet.

Large organizations, such as academic institutions, large enterprises, and governments, may perform the same function as ISPs, engaging in peering and purchasing transit on behalf of their internal networks. Research networks tend to interconnect with large sub networks such as GEANT, GLORIAD

**Modern uses**

- Communication

At the moment the easiest thing that can be done using the internet is that we can communicate with the people living far away from us with extreme ease. Earlier the communication used to be a daunting task, but all that changed once internet came into the life of the common people. Now people can not only chat but can also do the video conferencing. Communication is the most important gift that the internet has given to the common man.

- Research

In order to do research you need to go through hundreds of books as well as the references and that was one of the most difficult jobs to do earlier. Since the internet came into life, everything is available just a click away. You just have to search for the concerned topic and you will get hundreds of references that may be beneficial for your research.

- Financial Transaction

With the use of internet in the financial transaction, your work has become a lot easier. Now you don't need to stand in the queue at the branch of your particular bank, however you can just log in on to the bank website with the credential that has been provided to you by the bank and then can do any transaction related to finance at your will. With the ability to do the financial transaction easily over the internet you can purchase or sell items so easily. Financial transaction can be considered as one of the best uses of resource in the right direction.

- Leisure

Leisure is the option that we have next in the list. Right from watching your favourite videos to listening to songs, watching movies, playing games, chatting with the loved ones has been possible due to internet. Internet has progressed with so much pace that today whenever you get time, you just move on to the internet and enjoy such activities which help you relax. Leisure is one of the most important uses of internet and it has surely one thing that attracts people towards it.

- Shopping

Shopping has now become one of the most pleasing things to do using the internet.

Whenever you find time, just visit the concerned websites and order the items that you need from there. Those items will be delivered to you in best possible time. There is huge number of options available for a common person to buy or to sell any particular item using the internet. Using internet now makes it possibles to buy products from all over the world.

## *New words & Expressions:*

bar code    条形码
QR code    二维码
end-user    终端用户
copyright restriction    版权限制

## *Exercises*

1. Answer the following questions according to the text.

(1) What is the most common protocol found on the Internet and in home networks?

(2) How do the computers find each other in Internet?

(3) Please list the common use of the Internet.

2. Translate the following sentences into English.

(1) 现有的最大广域网是互联网。

(2) 通信是互联网给人们最重要的礼物。

## ☞ Section 4　Extended Reading:

# How to Use the Internet

Using the internet is a vital thing for this century. However, some people don't know how to use the web. Let us learn many of the ways that you can use the web.

### Part 1　Keeping up with friends and family

1. Use e-mail. E-mail is a lot like regular mail and you can use it in many of the same ways. You'll need to sign up with an e-mail service in order to get an address, though. Many e-mail services are free and good ones include GMail and Outlook.com. When you go to check your e-mail, you will go to the website for the service you signed up with and only the service you signed up with in order to read your mail.

E-mail addresses don't look like street addresses. They are in a format like yourname@website.com. For example, the e-mail to reach us here at wikiHow is wiki@wikihow.com. If your name is John Doe and you sign up with Gmail, your address could look like JohnDoe@gmail.com, JDoe@gmail.com, JohnD@gmail.com, or even something totally different like WarVet63@gmail.com.

2. Use Social Media. Social media is a term that encompasses a lot of different kinds of websites, all for connecting and communicating with other people. Commonly used forms of social media include:

• Facebook, which is used for lots of different purposes, from messaging to sharing pictures and videos.

• Twitter, which is used for sending very short updates and thoughts about your life.

- Instagram, which is used for sharing pictures.
- Pinterest, which is for sharing items you find on the internet.

3. Read or write blogs.

A blog, which was derived from the term web log, is an online journal. You can put texts, pictures, and even videos in a blog. You can write your own or you can read someone else's. Blogs cover all sorts of different subjects, and are starting to replace certain sections of the newspaper in their function.

4. Chat.

You can use the Internet to talk directly with people you know (or even people you don't know). If you want to talk face-to-face or with voice like a phone, you can use services like Skype, which is often free or a low price. You can also type-chat, which is like talking but with just text, using a number of different services (like AOL's Instant Messenger service AIM).

5. Start dating.

You can also date online! There are sites that are free, as well as sites that you pay for, all with the goal of helping you meet someone that's right for you. There are even specialty dating websites, for people in particular professions or with special interests. Match and eHarmony are the most common. Meetme is a popular free dating site and app for smartphones, tablets, and computers.

## Part 2  Keeping up with events

1. Read the news. You can read the newspaper online, often for free or cheaper than what you would pay for print. Most major newspapers include an online edition. These often may pair with videos for a multimedia experience. Try searching for your favorite newspaper! New York Times and CNN are common news sites.

2. Watch the news. You can also watch the news online. Go to the website for your local TV station to see what they offer, or watch clips from major news networks, like the BBC.

3. Get opinions and analysis. You can get op-ed style articles as well as financial, sports, and political analysis easily online, from blogs, news sites, and other websites. One popular

source of online analysis is Nate Silver, through his FiveThirtyEight blog.

4. Twitter. Twitter, while a common form of social media that is mainly used for telling all your friends about something weird you just ate, can also be used to keep up-to-date on important events. Follow Twitter feeds for official offices, like the White House or major news networks, to get the latest on events as they happen.

**Part 3   Managing your life**

1. Do online banking. Many major banks allow online banking, in which you can get your statements, make deposits and withdrawals, order checks, and do other common bank activities. Check the official site for your bank or call them to find out more information.

2. Pay your bills. You can also often pay your bills online or even set up automatic payments, so you don't have to worry about paying a particular bill each month. You can set this up through your bank's website (sometimes, depending on the bank) or you can go to the website for the company that you have to pay (if they have online bill-pay set up). Call them for more information.

3. Balance your checkbook. You can use free services like Google Spreadsheets to set up a tracker for your monthly expenses. This will be easier if you have experience with programs like Microsoft Excel, but you can also get templates that are easy to fill out. This service is free, as long as you have a Google Account.

4. Invest your money. If you love playing the stocks, you can even invest your money online, using websites like ETrade to buy, sell, and track your stocks. This is easy to do and gives you much more control over you trades.

5. Keep a calendar. You can keep a calendar with all of your appointments, birthdays, and anniversaries using tools like Google Calendar. You can even share your calendar with friends and family, so that they know where to find you and what's going on in your life.

6. Find a new job. If you want to get a paid job or even a volunteer position, you can find many opportunities online, using websites like Monster.com. You can search by what you want to do, where you are, how often you're available, etc. You can even do things like make resumes.

## Part 4  Researching information

1. Find professional services. The Internet is quickly becoming like a giant directory. Most professional services these days either have a website or at least a Google listing, so that you can easily get addresses and contact information, as well as hours and pricing. You can even use some websites to get recommendations, like AngiesList.com.

2. Take classes. You can take full university courses or even just free courses online, if you want to learn a new skill or keep your brain active. You can find free courses from major universities on websites like Coursera, but actual degree programs online usually cost money.

3. Learn new things. If you're in the mood for more of a lecture than a full class, you can also find this kind of bite-size info on the internet. Go to websites like TED to see interesting lectures from some of the best minds in the world, for free. You can learn loads of basic skills (and not-so-basic ones!) on websites like this one, wikiHow. You can also find websites like Wikipedia, which is a free online encyclopedia and contains a vast wealth of information. YouTube has lots of information and entertainment in video format.

4. Learn about your family history. If you're interested in your family history, you can do research about where you and your family come from. There are lots of ancestry websites which can provide not only information but also sometimes things like pictures or draft cards. Try Ancestry.com, FamilySearch.org, and EllisIsland.org. Many publicly available census records are also online.

## Part 5  Entertaining yourself

1. Watch television and movies. You don't have to have cable any more if you don't want to. Many popular TV shows and movies can be watched through services like Netflix or Hulu, which can even stream right to your TV. These usually cost money, but it's much, much cheaper than paying for cable.

2. Watch YouTube. Youtube carries all sorts of different video content. You can watch funny clips, family movies, full TV shows, full movies, clips of either of those things, or even just do things like listening to songs.

3. Play games. You can play games online (or even gamble!). Websites like Games.com

offer lots of free, classic games that you can play. Another option are games like fantasy football: a number of league are available online that you can enjoy for free.

4. Read comics. If you loved reading comics when they were in the newspaper, you can read many of those same comics online. Try a search for your favorite comic...you might be surprised!

- Read Garfield here.
- Read Family Circus here.
- Find new comics. There are lots of new comics that have never been up in the newspapers but can be read for free online. These are called webcomics, and cover a huge range of topics.

5. Listen to music. You can also listen to music online. There are many free websites that let you listen to music that you like. Pandora is a free Internet radio that lets you choose what kind of music to listen to. Slacker.com is similar to a satellite radio service with a wide variety of music of all genres. You can also try to bring up specific songs or artists using websites like YouTube.

## ☞ Section 5  Extended Reading:

## Set up an FTP between Two Computers

This article teaches you how to install, set up, and host an FTP server on your Windows computer. Once your FTP network is live, you can connect to it using another computer as long as you know the IP address of the FTP server. Unfortunately, macOS High Sierra removed FTP support, so you'll need to use a Windows computer for this process.

### Part 1  Preparing to create a server

1. Connect to a Wi-Fi network.

In order to host an FTP server, your computer must be connected to a wireless network.

2. Create a new folder somewhere.

If you don't already have a folder that you want to use as your computer's FTP storage, go to the location in which you want to create the folder (you can't use the This PC or Quick Access folder), then do the following:

- Right-click a blank space.
- Select New in the resulting drop-down menu.
- Click Folder in the New pop-out menu.
- Type in a name for your server, then press ↵ Enter.

3. Find out your router's IP address.

You'll need to know the IP address that your router uses in order to forward your FTP server's port later:

- Open Start.
- Click Network & Internet.
- Click Status.
- Click View your network properties.
- Scroll down to the "Wi-Fi" section.
- Look at the address to the right of the "Default Gateway" heading (it should be a collection of small numbers separated by periods).

**Part 2  Installing the server software**

1. Open Start.

Click the Windows logo in the bottom-left corner of the screen.

2. Open the Windows Features menu. Type in turn windows features on or off, then click Turn Windows features on or off when it appears at the top of the Start menu.

3. Expand the "Internet Information Services" heading.

Click the + icon to the left of this heading to do so.

4. Check the "FTP Server" box.

It's right of and below the "Internet Information Services" heading.

5. Expand the "FTP Server" menu.

Click the + icon to the left of "FTP Server" to do so.

6. Check the "FTP Extensibility" box. At this point, both of the "FTP Server" heading's boxes should be checked.

# Unit 6   Network and Internet

7. Check the "Web Management Tools" box.

This is the final box you need to check.

8. Click OK.

It's at the bottom of the window. Doing so will cause Windows to begin installing the necessary software.

The installation may take a few minutes to complete.

9. Click Close when prompted.

Now that your FTP server software is enabled, you can proceed with setting up the server itself.

**Part 3   Creating an FTP server**

1. Open Start.

Click the Windows logo in the bottom-left corner of the screen.

2. Open the IIS Manager.

Type in internet information services, then click Internet Information Services (IIS) Manager at the top of the Start menu.

3. Expand your computer's menu.

Click to the left of your computer's name in the top-left side of the window.

4. Right-click Sites.

A drop-down menu will appear.

5. Click Add FTP Site.

It's in the drop-down menu. Doing so opens a pop-up window.

6. Enter a site name.

Type whatever you want to name your FTP site into the top text box on this window.

7. Select your site's folder.

Click ... to the right of the bottom text box, then navigate to the location of the folder you want to use for your FTP server, click the folder, and click OK.

8. Click Next.

This is at the bottom of the page.

9. Check the "No SSL" box.

It's in the middle of the page. Since you're only using your FTP server to transfer files between two computers, it's okay to avoid using an SSL certificate here.

10. Click Next.

This is at the bottom of the page.

11. Check the "Basic" box.

It's near the top of the page. The "Basic" option will require users to log in using an e-mail address (or username).

12. Add yourself as the only user.

Click the "Allow access to" drop-down box, click Specific users, and then type your Windows 10 account's e-mail address into the text box below the drop-down box.

If you're using a local account with administrator privileges, you'll type in your account's username instead.

13. Check the "Read" and "Write" boxes.

Both are near the bottom of the window.

14. Click Finish.

It's at the bottom of the window. This will close the window, signifying that your FTP server has been created and turned on.

With these settings, your FTP server will go online whenever your computer is both on and connected to your Wi-Fi network.

**Part 4　Allowing the server in firewall**

1. Open Start.

Click the Windows logo in the bottom-left corner of the screen.

2. Open the "Allowed Apps" Firewall page.

Type in allow an app, then click Allow an app through Windows Firewall at the top of the Start window.

3. Click Change settings.

It's at the top of the window.

4. Scroll down to the "FTP Server" option.

This option is in the "F" section of allowed apps.

5. Check all three "FTP Server" boxes.

You'll see one box to the left of the "FTP server" heading as well as two boxes to the right; make sure that all three of these boxes have checkmarks in them.

6. Click OK.

It's at the bottom of the window. Your FTP server should now allow incoming connections.

**Part 5  Forwarding your router**

1. Open your router's page.

Type the router's address into the address bar of a web browser, then press Enter.

2. Log in if necessary.

If prompted for your login password and/or name, enter the credentials you created for your router.

If you're asked for credentials even though you didn't create any, look on the bottom of the router or check the router's manual for the default login credentials.

3. Set a static IP address for your computer.

In your router's list of connected items, find your computer's name and select the option to reserve or lock the IP address.

This option's location and appearance will vary depending on your router, so consult your router's documentation if you can't find the static IP section.

Doing this will ensure that your computer's IP address doesn't change. If you don't set a static IP address for your computer, you may lose connection to the FTP server later.

4. Make sure you know your computer's IP address.

In the static IP address section, note your computer's IP address. You'll need to know this number in order to add it to your forwarding rule.

5. Find and open the Port Forwarding section.

Again, this option's location will depend on your router, though you'll usually find it in the WAN, NAT, or Advanced section of settings.

6. Set both the inbound and outbound port to "21".

In both "Port" text boxes, make sure you enter the number 21.

7. Add your computer's static IP address.

It will usually go in the "Private IP" or "Address" box.

8. Select "TCP" as the forwarding rule.

Click the "UDP" or "TCP/UDP" drop-down box and then click TCP in the resulting drop-down menu. If you see TCP here already, skip this step.

9. Name the rule whatever you like.

Type whatever name you want to use (e.g., "FTP") into the "Name" text box.

10. Save and enable the rule.

Make sure the rule has a checkmark next to it (if possible), then click Save or OK to save the rule.

Your router may restart after approving these changes.

## Part 6  Connecting to the FTP server

1. Check your host computer's public IP address.

In order to connect to your FTP server from any location, you'll need to know the host computer's public IP address. On the computer on which you're running the FTP server, do the following:

- Go to https://www.google.com/.
- Type in what is my ip and press, Enter.
- Review the IP address at the top of the search results.

2. Open a web browser on the other computer.

On a computer other than the one on which you're hosting your FTP server, open any web browser other than Microsoft Edge.

3. Enter your FTP address.

Type ftp:// followed by the public IP address, then type: 21 at the end.

Your address should look something like this: ftp://123.456.78.901:21

4. Sign into the FTP server.

Enter your e-mail address, then click Sign In.

5. Add an FTP folder to your computer.

If you want to place a shortcut to the FTP server in your computer's This PC program, you can do so directly from within File Explorer:

- Open This PC.
- Click Computer, then click Add a network location.
- Click Next.
- Click Next twice.
- Enter your FTP server's address, then click Next.
- Uncheck the "Log on anonymously" box, then enter your FTP server's e-mail address into the "User name" box.
- Click Next, name the network if you like, click Next, and click Finish.

## ☞ Section 6  Extended Reading:

## To Be Completely Anonymous on the Internet

Internet is the place where we spent most of our time. We are now dependent on the internet for almost everything. We connect with friends and family through social networks, shop online, have fun and do serious works using the internet. However, internet is not a safe place anymore. Bad guys are trying to steal our identities and harm us. Sometimes our identities are being sold to other companies without our knowledge. When we visit any website our identities get exposed through IP address to the website owners. You should be very careful what you share on internet. Because once you share any sensitive information, it becomes almost impossible to completely delete it. Becoming completely anonymous on internet may not be possible but you can hide your identity to some extent by following some basic tips.

### 1. Hide IP address

Internet protocol address or IP address is our identity to the world when we are online. Every device connected to the internet has a unique IP address assigned by internet service providers. When we visit any website or access anything using internet, our IP address gets exposed. Some websites even store all details in their log files. IP address can reveal geographical location and some other information about us. We should hide our IP addresses in order to conceal our identities. There are several ways to hide IP address; you can either use proxy servers or use some software which will fake your IP address. Virtual Private Network or VPN like Cyber Ghost VPN can be used for this purpose as well.

**2. Fake your identity**

Yes, you read it right. If you want to be completely anonymous online, never use your actual name, date of birth or address on the internet. Instead, take a disguise; select a fake name, address and other related information when creating any online account. In this way you can safeguard your actual identity. Some sites like Google requires phone verification when opening a new account. I'll recommend you not to use these sites if you want to completely hide your personal information. However, keep in mind that providing incorrect personal information may have some legal issues.

**3. Stay away from social networks**

Social networks like Facebook, Twitter, Google+ have become a part of our daily life. It helps to stay connected with friends and family. But when it comes to identity theft, social networks are soft targets of hackers. Sometimes we share sensitive information and photos unknowingly. If you are already using social networks, then delete your all existing social network accounts. So if you want to be an anonymous, don't ever use any social networks.

**4. Don't use Google services**

You should not be using any Google services if you want to be anonymous. Don't perform a Google search or watch YouTube videos when you are logged into your Google account. Google stores all these activity and maintains a database of your account which can be accessed by visiting www.google.com/dashboard.

**5. Setup your own e-mail server or use disposable e-mail address**

If you have little technical knowledge you can buy a custom domain name and setup own e-mail service. We know that Google reads our e-mails for providing contextual advertisement. It is always a better idea to use own e-mail service rather than relying on any 3rd party e-mail service providers. Alternatively, you can use disposable e-mail address when signing up on websites that requires registration. In this way you can protect yourself from unwanted spam mails and hide your actual e-mail address.

**6. Disable cookies in browser and use private browsing mode**

Cookies are used for different purposes. It keeps a track of all websites visited, saves their log in credentials, etc. Advertising networks like Google AdSense serves targeted ads based on cookies. I'll recommend to completely disabling cookies. You can do that by going through options in any modern browser. Private browsing mode can be used to make our browsing details secret from others.

I think if you follow the above mentioned tips you can hide your identity. Don't forget to use good antivirus software and regularly update it. Enjoy your online life!

# Unit 7

# Information Security

After reading this unit and completing the exercises, you will be able to
- Understand the definition and key characteristics of information security.
- Be familiar with the history of information security.
- Use the knowledge of information security to protect the computer.
- Identify the importance of information technology.

## ☞ Section 1  Situational Dialogue:

### Computer Hackers

Tom: Hey, Mike. You've been surfing the Net for quite a while. What on earth are you searching for?

Mike: It's something relative to hackers. I often hear people talking about them, but I don't know much about them.

Tom: Well, roughly speaking, a hacker is a computer buff.

Mike: You mean a guy enthusiastic and knowledgeable about the computer?

Tom: You can say that.

Mike: But why are people always having such a negative attitude towards them?

Tom: They must have mixed hackers with crackers.

Mike: What is crackers then?

Tom: There is another group of people who loudly call themselves hackers, but they aren't. They break into computers and break the phone system. Real hackers call these people crackers, and want nothing to do with them.

Mike: So they are two totally different concepts.

Tom: Well, the real hackers mostly think crackers are lazy, irresponsible and not very bright, and feel that being able to break security does make you a hacker any more than being able to start cars without keys makes you an automotive engineer. Unfortunately, many journalists and writers have been fooled into using the word hacker to describe crackers. This irritates real hackers to no end.

Mike: I see. Then the basic difference is, hackers build things, crackers break them.

Tom: You got it.

Mike: Thanks a lot.

Tom: You are welcome.

## *New words & Expressions:*

hacker 黑客
enthusiastic 热情的
cracker 骇客
concept 概念

### ☞ Section 2 Reading Material:

# Computer Viruses

A computer virus is a computer program that can replicate itself and spread from one computer to another. The term "virus" is also commonly, but erroneously used, to refer to other types of malware, including but not limited to adware and spyware programs that do not have a reproductive ability.

Viruses can increase their chances of spreading to other computers by infecting files on a network file system or a file system that is accessed by other computers.

As stated above, the term "computer virus" is sometimes used as a catch-all phrase to include all types of malware, even those that do not have the ability to replicate themselves. Malware includes computer viruses, computer worms, Trojan horses, most rootkits, spyware, dishonest adware and other malicious or unwanted software, including true viruses. Viruses are sometimes confused with worms and Trojan horses, which are technically different. A worm can exploit security vulnerabilities to spread itself automatically to other computers through networks, while a Trojan horse is a program that appears harmless but hides malicious functions. Worms and Trojan horses, like viruses, may harm a computer system's data or performance. Some viruses and other malware have symptoms noticeable to the computer user, but many are surreptitious or simply do nothing to call attention to themselves. Some viruses do nothing beyond reproducing themselves.

An example of a virus which is not a malware, but is putatively benevolent is Fred Cohen's compression virus. However, antivirus professionals do not accept the concept of benevolent viruses, as any desired function can be implemented without involving a virus (automatic compression, for instance, is available under the Windows operating system at the choice of the user). Any virus will by definition make unauthorised changes to a computer, which is undesirable even if no damage is done or intended.

The Creeper virus was first detected on ARPANET, the forerunner of the Internet, in the early 1970s. Creeper was an experimental self-replicating program written by Bob Thomas at BBN Technologies in 1971. Creeper used the ARPANET to infect DEC PDP-10 computers running the TENEX operating system. Creeper gained access via the ARPANET and copied itself to the remote system where the message, "I'm the creeper, catch me if you can!" was displayed. The Reaper program was created to delete Creeper.

A program called "Elk Cloner" was the first personal computer virus to appear "in the wild"—that is, outside the single computer or lab where it was created. Written in 1981 by Richard Skrenta, it attached itself to the Apple DOS 3.3 operating system and spread via floppy disk. This virus, created as a practical joke when Skrenta was still in high school, was injected in a game on a floppy disk. On its 50th use the Elk Cloner virus would be activated, infecting the personal computer and displaying a short poem beginning "Elk Cloner: The program with a personality."

The first IBM PC virus in the wild was a boot sector virus dubbed (c)Brain, created in 1986 by the Farooq Alvi Brothers in Lahore, Pakistan, reportedly to deter piracy of the software they had written.

Before computer networks became widespread, most viruses spread on removable media, particularly floppy disks. In the early days of the personal computer, many users regularly exchanged information and programs on floppies. Some viruses spread by infecting programs stored on these disks, while others installed themselves into the disk boot sector, ensuring that they would be run when the user booted the computer from the disk, usually inadvertently. Personal computers of the era would attempt to boot first from a floppy if one had been left in the drive. Until floppy disks fell out of use, this was the most successful infection strategy and boot sector viruses were the most common in the wild for many years.

Traditional computer viruses emerged in the 1980s, driven by the spread of personal computers and the resultant increase in BBS, modem use, and software sharing. Bulletin board-driven software sharing contributed directly to the spread of Trojan horse programs, and viruses were written to infect popularly traded software. Shareware and bootleg software were equally common vectors for viruses on BBSs.

Macro viruses have become common since the mid-1990s. Most of these viruses are written in the scripting languages for Microsoft programs such as Word and Excel and spread throughout Microsoft Office by infecting documents and spreadsheets. Since Word and Excel were also available for Mac OS, most could also spread to Macintosh computers. Although most of these viruses did not have the ability to send infected email messages, those viruses did take advantage of the Microsoft Outlook COM interface.

## *New words & Expressions:*

| | |
|---|---|
| hacking program | 黑客程序 |
| spyware | 间谍软件，间谍程序 |
| worm | 蠕虫病毒 |
| antivirus program | 防病毒程序，杀毒软件 |
| virus | 病毒 |

| | |
|---|---|
| firewall | 防火墙 |
| adware | 恶意广告软件 |
| malware | 恶意软件 |
| Trojan horse | 特洛伊木马 |
| vulnerability | 弱点，攻击 |
| creeper | 爬虫(病毒) |
| shareware | 共享软件 |
| bootleg | 非法制作或者贩卖的，盗版的 |

## *Exercises*

1. Answer the following questions according to the text.

(1) What is computer virus?

(2) When did traditional computer viruses emerge?

(3) When were viruses that spread using cross-site scripting first reported?

(4) In what way can viruses increase their chances of spreading to other computers on a network file system or a file system that is accessed by other computers?

2. Translate the following sentences into Chinese.

(1) A computer virus is a computer program that can replicate itself and spread from one computer to another.

(2) Viruses can increase their chances of spreading to other computers by infecting files on a network file system or a file system that is accessed by other computers.

3. Match the items in Column A with the translated versions in Column B.

A
a. 蠕虫
b. 数据加密
c. 防火墙
d. 黑客
e. 数据库
f. 恶意软件

B
(　) (1) hacker
(　) (2) data encryption
(　) (3) database
(　) (4) worm
(　) (5) fi rewall
(　) (6) malware

4. Fill in the blanks with the right words.

> ARPANET, replicate, infecting, programs, Macro viruses, personal, computer

(1) The Creeper virus was first detected on _____, the forerunner of the Internet, in the early 1970s.

(2) The term "computer virus" is sometimes used as a catch-all phrase to include all types of malware, even those that do not have the ability to _____ themselves.

(3) Some viruses spread by _____ stored on these disks, while others installed themselves into the disk boot sector, ensuring that they would be run when the user booted the computer from the disk, usually inadvertently.

(4) Most of _____ are written in the scripting languages for Microsoft programs such as Word and Excel and spread throughout Microsoft Office by infecting documents and spreadsheets.

(5) A program called "Elk Cloner" was the first _____ virus to appear "in the wild" —that is, outside the single computer or lab where it was created.

## ☞ Section 3  Reading Material:

# The Ways to Protect Information Security

Information security is the process of protecting the availability, privacy, and integrity of data. While the term often describes the measures and methods of increasing computer security, it also refers to the protection of any type of important data, such as personal diaries or the classified plot details of an upcoming book. No security system is foolproof, but taking basic and practical steps to protect data is critical for good information security.

**Password protection**

Using passwords is one of the most basic methods of improving information security. This measure reduces the number of people who have easy access to the information, since only those with approved codes can reach it. Unfortunately, passwords are not foolproof, and hacking programs can run through millions of possible codes in just seconds. Passwords can also be breached through carelessness, such as by leaving a public computer logged into an account or using a simple code, like "password" or "1234".

**Antivirus and malware protection**

One way that hackers gain access to secure information is through malware, which includes computer viruses, spyware, worms, and other programs. These pieces of code are installed on computers to steal information, limit usability, record user actions, or destroy data. Using strong antivirus software is one of the best ways of improving information security. Antivirus programs scan the system to check for any known malicious software, and most programs will warn the user if he or she is on a webpage that contains a potential virus. Most programs will also perform a scan of the entire system on command, identifying and destroying any harmful objects.

**Firewalls**

A firewall helps maintain computer information security by preventing unauthorized access to a network. There are several ways to do this, including by limiting the types of data allowed in and out of the network, re-routing network information through a proxy server to hide the real address of the computer, or by monitoring the characteristics of the data to determine if it's trustworthy. In essence, firewalls filter the information that passes through them, only allowing authorized content in. Specific websites, protocols (like File Transfer Protocol or FTP), and even words can be blocked from coming in, as can outside access to computers within the firewall.

**Legal liability**

Businesses and industries can also maintain information security by using privacy laws. Workers at a company that handle secure data may be required to sign Non-Disclosure Agreements(NDAs), which forbid them from revealing or discussing any classified topics. If an employee attempts to give or sell secrets to a competitor or other unapproved source, the company can use the NDAs as grounds for legal proceedings. The use of liability laws can help companies preserve their trademarks, internal processes, and research with some degree of reliability.

**Training and common sense**

One of the greatest dangers to computer data security is human error or ignorance. Those responsible for using or running a computer network must be carefully trained in order to avoid accidentally opening the system to hackers. In the workplace, creating a training program that includes information on existing security measures as well as permitted and prohibited computer usage can reduce breaches in internal security. Family members on a home network should be taught about running virus scans, identifying potential Internet threats, and protecting personal information online.

## *New words & Expressions:*

| | |
|---|---|
| integrity | 完整性 |
| reduce | 降低，减少 |
| not foolproof | 不是万无一失 |

## *Exercises*

1. Answer the following questions according to the text.

(1) What aspects does information guarantee include?

(2) How to set up password ?

(3) What is the function of the firewall?

(4) What is the malware?

2. Translate the following sentences into English/Chinese.

(1) 使用密码是提高信息安全的最基本的方法之一。

(2) A firewall helps maintain computer information security by preventing unauthorized access to a network.

(3) Antivirus programs scan the system to check for any known malicious software, and most programs will warn the user if he or she is on a webpage that contains a potential virus.

## ☞ Section 4  Extended Reading:

## What Does a Data Security Manager Do?

The data security manager is responsible for the oversight of business applications where sensitive data are stored or transmitted. His job is to protect the personal information of both employees and customers by implementing and maintaining necessary internal security functions. This individual acts as a consultant on all business processes that require security features.

This job calls for the documentation of security policies and procedures to make sure they meet industry standards. The data security manager provides training to employees on how to properly use security functions to protect their private data. He makes sure that special security clearance is given to the correct individuals and that the appropriate privileges have been granted. This person facilitates internal meetings to promote good security practices and to offer updates on security enhancements.

The data security manager is tasked with performing assessments on security risks to the company. He audits all security functions and produces reports offering suggestions or comments. This employee is in frequent contact with higher-level managers in the company. As new threats emerge, this person is constantly offering new solutions and recommendations for necessary tweaks to the security systems.

This individual is typically the person to tailor specific security policies for the company and ensure that they are carried out. Departments installing new data systems are usually required to contact the data security manager with their plan on maintaining security within the new system. The security manager will consult with the department on the installation and implementation of the new system so that data remains secured. He is also responsible for preparing the company for a significant security breach by establishing procedures and guidelines for handling such a situation.

The position of data security manager often requires someone knowledgeable in both electronic and physical security functions. He must understand how to secure internal networks from hackers and viruses. Designing and managing an effective system of firewalls is an essential task for this employee. Making sure all pertinent data are encrypted across the network and is made available only to authorized persons is another important job of the security manager. In addition, he must be able to install, manage, and maintain physical security measures, such as key card and fingerprint authorizations systems.

The data security manager keeps daily inventory of new security threats to internal networks and data systems. He checks to ensure that software is updated and patched to fix all security holes. This individual must have the ability to establish working relationships with all employees and answer internal security questions. The safety and security of the entire organization is in the hands of this individual.

## ☞ Section 5　Extended Reading:

# How to Create a Secure and Stable Windows System

A secure and stable system is essential to every computer user. How can we possess such a system? The following will teach how to build your own security systems.

Try to install the operating system in English version.

If you just want to improve the security of your operating system, I recommend that you install the original Windows' English version. Because when a new vulnerability is discovered, the patches in English version usually act the first, while other versions come after a span. And this spacing interval may decide the result of the system.

**Undo useless components.**

When Windows system is installed, it will prompt us to install some components. In general, the components are unnecessary you could ignore them. For ordinary users, there is no need to install Windows 2000/XP's Internet Information Services (IIS), so that they can naturally avoid some external attacks through IIS by PRINTER, IDQ, IDA, or WEBDEV.

**Select secure file format.**

For the Windows 2000/XP users, NTFS file format may be their best choice. Because no matter from the speed of file retrieval or the access control of system resources, NTFS is significantly better than the FAT system. We can right click on the disk partition that uses the NTFS format and select "Properties" on pop-up menu. Then we will see the spare "quota" and "security" on the disk in NTFS format. Through the two tabs the users can detailedly set the access right to the logic disk.

**Have system services custom-made.**

Windows 2000/XP system will provide users with many services after a normal start while all of which are not needed by most users. Obviously, extra services can only increase the load and instability of system. On the desktop, we can right-click "My Computer ? Management", and then in the left side of the interface window that has opened, select "Services and Applications? Services", where we can turn of the unnecessary services to improve system's stability, security and speed up system speed. It is emphasized that services such as Remote Registry Service and Telnet must be stopped: Double-click the relevant project, and set them to "Manual" or "No" in the open window.

### ☞ Section 6  Extended Reading:

## Create a Perfect Password: Ten Easy Steps to Stay Secure

If you're one of the millions of people whose password to their online accounts is "password", don't feel bad — you're not alone. Remembering a single PIN, password, or secret phrase can sometimes be bothersome — let alone passwords for the dozens of accounts and devices many people have nowadays.

Online-security experts recommend long, strong passwords for a reason—identity and information theft are rampant, and hackers have many tools at their disposal that allow them to crack simple passwords like "123456" and "abcdefg". In order to protect your identity and online information, a tougher password is a must. But there's no need to memorize hexadecimal strings of random characters; there are several easy ways to create — and remember — strong, safe passwords.

**Go for length.**

The best passwords are at least seven characters long, and hopefully as long as fourteen characters. The shorter a password is, the easier it is to crack.

**Find something random.**

Instead of using a word as your password, use a favorite quote, lyric, or phrase (containing at least ten words), and use the first letter of each word as your password. If you're going to San Francisco, be sure to wear some flowers in your hair becomes "iygtsfbstwsfiyh". Although the sequence is memorable and makes sense to you, it seems random to anyone else.

Another way to find a random password is to use an online password-generator service, such as StrongPasswordGenerator.com, and then create a mnemonic device to help you remember it. When the service supplies a random sequence like "Jni8e8r," remember it by teaching yourself the phrase "Jeffrey normally inspired eighty-eight rainbows."

**Misspel deliberately.**

This doesn't mean using common misspellings of regular words; rather, devise a creative misspelling of a word you can remember and that can make your password safer. For example, "Paris" can become "Pearisse."

**Add some complexity.**

Good passwords contain symbols, punctuations, deliberate misspellings, and a blend of lowercase and capital letters. Turn a simple password like "catlover" into a more secure version like "c@LUVr"!

**Add numbers.**

Passwords with numbers are harder to crack, but don't use easy-to-guess numbers, such as the current year or your birthday. Choose seemingly random numbers (that have significance to you) and place them in the middle of the text for maximum security, or substitute numbers for multiple letters. An easy password like "basketball" can become "8a5k3tba1l".

**Mix it up.**

The very best passwords use a blend of all these techniques, so be sure to employ at least two or three to create the most powerful protection. If you have a favorite phrase that you've distilled to an acronym, add some capital letters or punctuation. Add length to a short password with numbers, and add complexity to a deliberate misspelling with characters or symbols. Using a variety of password-enhancing tricks ensures a better result.

**Check it out.**

Use a password checker to make sure that your password is as strong as it can be. If your password is rated weak or medium, you may want to add more numbers, symbols, or other characters to make it longer and more complex.

Use a different password for each account.

As maddening as it can be to keep track of all those passwords, it really is important not to use the same password for every online account, or a hacker who gains passwords from one site can use your e-mail address to compromise any other accounts you hold. But if you must reuse passwords, at the very least you should use separate passwords for banking or financial accounts, and leave the weaker or all-purpose passwords for accounts without access to your financial data. One way to keep track of multiple passwords is by using a password manager. These online or USB-based encrypted programs, like Password Dragon and KeePass, store passwords for all your accounts, leaving you to remember only one master password.

**Change them often.**

Computer hacking algorithms get more sophisticated every day, so it's best to change your passwords every few months, if not sooner. This is especially true for your password to any site that stores financial information, since these are more likely to be the target of a cyber-attack.

**Don't make it easy.**

• A surprising number of people use simple passwords like "password", which are incredibly easy for a computer to decipher or a person to guess. When formulating your passwords, never rely on these usual suspects.

• Personal information like your name, birthdate, address, phone number, or license plate number.

• Repetitive sequences, like "121212" or "bbbbbb," or adjacent letters on the keyboard, like "qwerty" or "asdfgh".

- Real words that appear in the dictionary, including common misspellings of those words.
- Real words spelled in reverse ("drowssap").
- Real words with a single number at the end ("password4").
- A real word with one letter replaced by a number ("passw0rd").

Although it's common to use a simple password like "password" or "123456", it's an invitation to disaster. Creating a better password takes only a few moments, and keeping your information safe is well worth the effort.

# Appendix

## Index of Basic Vocabulary

| | |
|---|---|
| advanced application | 高级应用 |
| analytical graph | 分析图表 |
| analyze | 分析 |
| animation | 动画 |
| application software | 应用软件 |
| arithmetic operation | 算术运算 |
| audio-output device | 音频输出设备 |
| access time | 存取时间 |
| access | 存取 |
| add-ons | 插件 |
| address | 地址 |
| agent | 代理 |
| analog signal | 模拟信号 |
| applet | 程序 |
| asynchronous communications port | 异步通信端口 |
| attachment | 附件 |

| | |
|---|---|
| bar code | 条形码 |
| bar code reader | 条形码读卡器 |

| | |
|---|---|
| basic application | 基础程序 |
| binary system | 二进制系统 |
| bit | 比特 |
| browser | 浏览器 |
| bus line | 总线 |
| bandwidth | 带宽 |
| bluetooth | 蓝牙 |
| broadband | 宽带 |
| business-to-business | 企业对企业电子商务 |
| business-to-consumer | 企业对消费者 |
| bus | 总线 |

| | |
|---|---|
| cable | 连线 |
| cable modem | 有线调制解调器 |
| CD-ROM | 可记录光盘 |
| CD-RW | 可重写光盘 |
| CD-R | 可记录压缩光盘 |
| cell | 单元箱 |
| chain printer | 链式打印机 |
| character and recognition device | 字符标识识别设备 |
| chart | 图表 |
| chassis | 支架 |
| chip | 芯片 |
| clarity | 清晰度 |
| column | 列 |
| combination key | 结合键 |
| computer competency | 计算机能力 |

| | |
|---|---|
| connectivity | 连接，结点 |
| control unit | 操纵单元 |
| cordless or wireless mouse | 无线鼠标 |
| channel | 信道 |
| chat group | 谈话群组 |
| chlorofluorocarbons(CFCs) | 氯氟甲烷 |
| client | 客户端 |
| coaxial cable | 同轴电缆 |
| cold site | 冷战 |
| commerce server | 商业服务器 |
| communication channel | 信道 |
| communication system | 信息系统 |
| compact disc rewritable | 可擦写光盘 |
| compact disc | 光盘 |
| computer crime | 计算机犯罪 |
| computer ethics | 计算机道德 |
| computer network | 计算机网络 |
| computer support specialist | 计算机支持专家 |
| computer technician | 计算机技术人员 |
| computer trainer | 计算机教师 |
| connection device | 连接设备 |
| connectivity | 连接 |
| consumer-to-consumer | 个人对个人 |
| cookies-cutter program | 信息记录截取程序 |
| cookie | 信息记录程序 |
| cracker | 解密高手 |
| cybercash | 电子现金 |
| cyberspace | 计算机空间 |

| | |
|---|---|
| database | 数据库 |
| database file | 数据库文件 |
| database manager | 数据库管理 |
| data bus | 数据总线 |
| data projector | 数码放映机 |
| desktop system unit | 台式电脑系统单元 |
| destination file | 目标文件 |
| digital camera | 数码照相机 |
| digital notebook | 数字笔记本 |
| digital bideo camera | 数码摄影机 |
| document | 文档 |
| document file | 文档文件 |
| dot-matrix printer | 点矩阵式打印机 |
| dual-scan monitor | 双向扫描显示器 |
| dumb terminal | 非智能终端 |
| data security | 数据安全 |
| database administrator | 数据库管理员 |
| dataplay | 数字播放器 |
| demodulation | 解调 |
| denial of service attack | 拒绝服务攻击 |
| dial-up service | 拨号服务 |
| digital cash | 数字现金 |
| digital signal | 数字信号 |
| digital subscriber line | 数字用户线路 |
| digital versatile disc | 数字化通用磁盘 |
| digital video disc | 数字化视频光盘 |

| | |
|---|---|
| direct access | 直接存取 |
| directory search | 目录搜索 |
| disaster recovery plan | 灾难恢复计划 |
| disk caching | 磁盘驱动器高速缓存 |
| diskette | 磁盘 |
| disk | 磁碟 |
| distributed data processing system | 分部数据处理系统 |
| distributed processing | 分布处理 |
| domain code | 域代码 |
| downloading | 下载 |
| DVD | 数字化通用磁盘 |
| DVD-R | 可写 DVD |
| DVD-RAM DVD | 随机存取器 |
| DVD-ROM | 只读 DVD |

| | |
|---|---|
| e-book | 电子阅读器 |
| end user | 终端用户 |
| e-cash | 电子现金 |
| e-commerce | 电子商务 |
| electronic cash | 电子现金 |
| electronic commerce | 电子商务 |
| encrypting | 加密术 |
| energy star | 能源之星 |
| enterprise computing | 企业计算化 |
| environment | 环境 |
| erasable optical disk | 可擦除式光盘 |
| external modem | 外置调制解调器 |

| | |
|---|---|
| expansion card | 扩展卡 |
| extranet | 企业外部网 |

| | |
|---|---|
| fax machine | 传真机 |
| field | 域 |
| find | 搜索 |
| firewire port port | 火线端口 |
| firmware | 固件 |
| flash RAM | 闪存 |
| flatbed scanner | 台式扫描器 |
| flat-panel monitor | 纯平显示器 |
| floppy disk | 软盘 |
| formatting toolbar | 格式化工具条 |
| formula | 公式 |
| function | 函数 |
| fiber-optic cable | 光纤电缆 |
| file compression | 文件压缩 |
| file decompression | 文件解压缩 |
| filter | 过滤 |
| firewall | 防火墙 |
| fixed disk | 固定硬盘 |
| flash memory | 闪存 |
| flexible disk | 可折叠磁盘 |
| floppies | 磁盘 |
| floppy disk | 软盘 |
| floppy-disk cartridge | 磁盘盒 |

| formatting | 格式化 |
| full-duplex communication | 全双通通信 |

| general-purpose application | 通用运用程序 |
| gigahertz | 千兆赫 |
| graphic tablet | 绘图板 |
| green pc | 绿色个人计算机 |

| handheld computer | 手提电脑 |
| hard copy | 硬拷贝 |
| hard disk | 硬盘 |
| hardware | 硬件 |
| help | 帮助 |
| host computer | 主机 |
| home page | 主页 |
| hyperlink | 超链接 |
| hacker | 黑客 |
| half-duplex communication | 半双通通信 |
| hard disk | 硬盘 |
| hard-disk cartridge | 硬盘盒 |
| hard-disk pack | 硬盘组 |
| head crash | 磁头碰撞 |
| header | 标题 |
| help desk specialist | 帮助办公专家 |
| helper application | 帮助软件 |
| hierarchical network | 层次型网络 |

| | |
|---|---|
| history file | 历史文件 |
| hits | 匹配记录 |
| horizontal portal | 横向用户 |
| hot site | 热战 |
| hybrid network | 混合网络 |
| hyperlink | 超链接 |

| | |
|---|---|
| i-drive | 网络硬盘驱动器 |
| image capturing device | 图像获取设备 |
| information technology | 信息技术 |
| ink-jet printer | 墨水喷射印刷机 |
| integrated package | 综合性组件 |
| intelligent terminal | 智能终端设备 |
| intergrated circuit | 集成电路 |
| interface card | 接口卡 |
| internal modem | 内部调制解调器 |
| internet telephony | 网络电话 |
| internet terminal | 互联网终端 |
| Identification | 识别 |
| index search | 索引搜索 |
| information pusher | 信息推送器 |
| initializing | 初始化 |
| instant messaging | 计时信息 |
| internal hard disk | 内置硬盘 |
| Internal modem | 内部调制解调器 |
| Internet hard drive | 网络硬盘驱动器 |
| Intranet | 企业内部网 |

## J

| | |
|---|---|
| Joystick | 操纵杆 |

## K

| | |
|---|---|
| keyword search | 关键字搜索 |

## L

| | |
|---|---|
| laser printer | 激光打印机 |
| layout file | 版式文件 |
| light pen | 光笔 |
| locate | 定位 |
| logical operation | 逻辑运算 |
| land | 凸面 |
| line of sight communication | 视影通信 |
| low bandwidth | 低带宽 |
| lurking | 潜伏 |

## M

| | |
|---|---|
| main board | 主板 |
| mark sensing | 标志检测 |
| mechanical mouse | 机械鼠标 |
| memory | 内存 |
| menu | 菜单 |
| menu bar | 菜单条 |
| microprocessor | 微处理器 |
| microsecond | 微秒 |

| | |
|---|---|
| modem card | 调制解调器 |
| monitor | 显示器 |
| motherboard | 主板 |
| mouse | 鼠标 |
| multifunctional device | 多功能设备 |
| magnetic tape reel | 磁带卷 |
| magnetic tape streamer | 磁带条 |
| mailing list | 邮件列表 |
| medium band | 媒质带宽 |
| metasearch engine | 整合搜索引擎 |
| microwave | 微波 |
| modem | 解调器 |
| modulation | 解调 |

## N

| | |
|---|---|
| net PC | 网络计算机 |
| network adapter card | 网卡 |
| network personal computer | 网络个人电脑 |
| network terminal | 网络终端 |
| notebook computer | 笔记本电脑 |
| notebook system unit | 笔记本系统单元 |
| numeric entry | 数字输入 |
| network architecture | 网络体系结构 |
| network bridge | 网桥 |
| network gateway | 网关 |
| network manager | 网络管理员 |
| newsgroup | 新闻组 |
| node | 节点 |
| nonvolatile storage | 非易失性存储 |

## O

| | |
|---|---|
| object embedding | 对象嵌入 |
| object linking | 目标链接 |
| open architecture | 开放式体系结构 |
| optical disk | 光盘 |
| optical mouse | 光电鼠标 |
| optical scanner | 光电扫描仪 |
| outline | 大纲 |
| off-line browser | 离线浏览器 |
| online storage | 联机存储 |

## P

| | |
|---|---|
| palmtop computer | 掌上电脑 |
| parallel ports | 并行端口 |
| passive-matrix | 被动矩阵 |
| PC card | 个人计算机卡 |
| personal laser printer | 个人激光打印机 |
| personal video recorder card | 个人视频记录卡 |
| photo printer | 照片打印机 |
| pixel | 像素 |
| platform scanner | 平版式扫描仪 |
| plotter | 绘图仪 |
| plug and play | 即插即用 |
| plug-in board | 插件卡 |
| pointer | 指示器 |
| pointing stick | 指示棍 |
| port | 端口 |

| | |
|---|---|
| portable scanner | 便携式扫描仪 |
| presentation file | 演示文稿 |
| presentation graphic | 电子文稿程序 |
| primary storage | 主存 |
| procedure | 规程 |
| processor | 处理机 |
| programming control language | 程序控制语言 |
| packet | 数据包 |
| parallel data transmission | 平行数据传输 |
| peer-to-peer network system | 得等网络系统 |
| person-person auction site | 个人对个人拍卖站点 |
| physical security | 物理安全 |
| pit | 凹面 |
| plug-in | 插件程序 |
| polling | 轮询 |
| privacy | 隐私权 |
| proactive | 主动地 |
| programmer | 程序员 |
| protocol | 协议 |
| provider | 供应商 |
| proxy server | 代理服务 |
| pull product | 推取程序 |
| push product | 推送程序 |

| | |
|---|---|
| RAM cache | 随机高速缓冲器 |
| range | 范围 |
| record | 记录 |

| | |
|---|---|
| relational database | 关系数据库 |
| replace | 替换 |
| resolution | 分辨率 |
| row | 行 |
| read-only | 只读 |
| reformatting | 重组 |
| regional service provider | 区域性服务供应商 |
| repetitive motion injury | 反复性动作损伤 |
| reverse directory | 反向目录 |
| ring network | 环形网 |

## S

| | |
|---|---|
| scanner | 扫描仪 |
| search | 查找 |
| secondary storage device | 助存储设备 |
| semiconductor | 半导体 |
| serial ports | 串行端口 |
| server | 服务器 |
| shared laser printer | 共享激光打印机 |
| sheet | 表格 |
| silicon chip | 硅片 |
| slot | 插槽 |
| smart card | 智能卡 |
| soft copy | 软拷贝 |
| software suite | 软件协议 |
| sorting | 排序分类 |
| source file | 源文件 |
| special-purpose application | 专用文件 |
| spreadsheet | 电子数据表 |

| | |
|---|---|
| standard toolbar | 标准工具栏 |
| supercomputer | 巨型机 |
| system cabine | 系统箱 |
| system clock | 时钟 |
| system software | 系统软件 |
| satellite/air connection service | 卫星无线连接服务 |
| search engine | 搜索引擎 |
| search provider | 搜索供应者 |
| search service | 搜索服务器 |
| sector | 扇区 |
| security | 安全 |
| sending and receiving device | 发送接收设备 |
| sequential access | 顺序存取 |
| serial data transmission | 单向通信 |
| signature line | 签名档 |
| snoopware | 监控软件 |
| software piracy | 软件盗版 |
| solid-state storage | 固态存储器 |
| specialized search engine | 专用搜索引擎 |
| spider | 网页爬虫 |
| spike | 尖峰电压 |
| star network | 星型网 |
| strategy | 方案 |
| subject | 主题 |
| subscription address | 预定地址 |
| superdisk | 超级磁盘 |
| surfing | 网上冲浪 |
| surge protector | 浪涌保护器 |
| systems analyst | 系统分析师 |

| | |
|---|---|
| table | 二维表 |
| telephony | 电话学 |
| television board | 电视扩展卡 |
| terminal | 终端 |
| template | 模板 |
| text entry | 文本输入 |
| thermal printer | 热印刷 |
| thin client | 瘦客 |
| toggle key | 触发键 |
| toolbar | 工具栏 |
| touch screen | 触摸屏 |
| trackball | 追踪球 |
| TV tuner card | 电视调谐卡 |
| two-state system | 双状态系统 |
| technical writer | 技术协作者 |
| technostress | 重压技术 |
| telnet | 远程登录 |
| time-sharing system | 分时系统 |
| topology | 拓扑结构 |
| track | 磁道 |
| traditional cookie | 传统的信息记录程序 |
| twisted pair | 双绞 |

| | |
|---|---|
| unicode | 统一字符标准 |
| uploading | 上传 |

usenet 世界性新闻组网络

| | |
|---|---|
| virtual memory | 虚拟内存 |
| video display screen | 视频显示屏 |
| voice recognition system | 声音识别系统 |
| vertical portal | 纵向门户 |
| virus checker | 病毒检测程序 |
| virus | 病毒 |
| voiceband | 音频带宽 |
| volatile storage | 易失性存储 |
| voltage surge | 冲击性电压 |

| | |
|---|---|
| wand reader | 条形码读入 |
| web | 网络 |
| web appliance | 环球网设备 |
| web page | 网页 |
| web site address | 网络地址 |
| web terminal | 环球网终端 |
| webcam | 摄像头 |
| what-if analysis | 假定分析 |
| wireless revolution | 无线革命 |
| word | 字长 |
| word processing | 文字处理 |
| word wrap | 自动换行 |
| worksheet file | 工作表文件 |
| web auction | 网上拍卖 |

| | |
|---|---|
| web broadcaste | 网络广播 |
| web portal | 门户网站 |
| web site | 网站 |
| web storefront | 网上商店 |
| web utilitie | 网上应用程序 |
| web-downloading utilitie | 网页下载应用程序 |
| webmaster web | 站点管理员 |
| wireless modem | 无线调制解调器 |
| wireless service provider | 无线服务供应商 |
| world wide web | 万维网 |
| worm | 蠕虫病毒 |
| write-protect notch | 写保护口 |

## Terminology

| | |
|---|---|
| DVD (Digital Video Disc) | 数字视频光盘 |
| IT (Information Technology) | 信息技术 |
| CD (Compact Disc) | 压缩盘 |
| PDA (Personal Digital Assistant) | 个人数字助理 |
| RAM (Random Access Memory) | 随机存储器 |
| WWW (World Wide Web) | 万维网 |
| DBMS (Database Management System) | 数据库管理系统 |
| HTML (Hypertext Markup Language) | 超文本标示语言 |
| OLE (Object Linking and Embedding) | 对象链接入 |
| SQL (Structured Query Language) | 结构化查询语言 |
| URL (Uniform Resource Locator) | 统一资源定位器 |
| AGP (Accelerated Graphics Port) | 加速图形接口 |
| ALU (Arithmetic-Logic Unit) | 算术逻辑单元 |

| | |
|---|---|
| CPU (Central Processing Unit) | 中央处理器 |
| CMOS (Complementary Metal-Oxide Semiconductor) | 互补金属氧化物半导体 |
| CISC (Complex Instruction Set Computer) | 复杂指令集计算机 |
| HPSB (High Performance Serial Bus) | 高性能串行总线 |
| ISA (Industry Standard Architecture) | 工业标准结构体系 |
| PCI (Peripheral Component Interconnect) | 外部设备互连总线 |
| PCMCIA (Personal Computer Memory Card International Association) | 个人计算机存储卡国际协会 |
| RAM (Random-Access Memory) | 随机存储器 |
| ROM (Read-Only Memory) | 只读存储器 |
| USB (Universal Serial Bus) | 通用串行总线 |
| CRT (Cathode-Ray Tube) | 阴极射线管 |
| HDTV (High-Definition Television) | 高清晰度电视 |
| LCD (Liquid Crystal Display monitor) | 液晶显示器 |
| MICR (Magnetic-Ink Character Recognition) | 磁墨水字符识别器 |
| OCR (Optical-Character Recognition) | 光电字符识别器 |
| OMR (Optical-Mark Recognition) | 光标阅读器 |
| TFT (Thin Film Transistor monitor) | 薄膜晶体管显示器 |
| Zip disk | 压缩磁盘 |
| DNS (Domain Name System) | 域名服务器 |
| FTP (File Transfer Protocol) | 文件传送协议 |
| HTML (Hypertext Markup Language) | 超文本链接标识语言 |
| LAN (Local Area Network) | 局域网 |
| IRC (Internet Relay Chat) | 互联网多线交谈 |
| MAN (Metropolitan Area Network) | 城域网 |
| NOS (Network Operation System) | 网络操作系统 |
| URL (Uniform Resource Locator) | 统一资源定位器 |
| WAN (Wide Area Network) | 广域网 |

# References

[1] 陈贤平. 计算机专业英语. 北京：清华大学出版社，2010.
[2] 郭敏. 计算机专业英语. 北京：中国水利水电出版社，2011.
[3] 王小刚. 计算机专业英语. 北京：机械工业出版社，2010.